Biochemistry, Molecular Biology and Biotechnology

NIPA® GENX ELECTRONIC RESOURCES & SOLUTIONS P. LTD.
New Delhi-110 034

Biochemistry, Molecular Biology and Biotechnology

– Instant Notes

By :
H. P. Gajera (Associate Professor)
S. V. Patel (Professor)
B. A. Golakiya (Professor & Head)
Department of Biotechnology
College of Agriculture
Junagadh Agricultural University
Junagadh, Gujarat - 362 001

NIPA® GENX ELECTRONIC RESOURCES & SOLUTIONS P. LTD.
New Delhi-110 034

NIPA® GENX ELECTRONIC
RESOURCES & SOLUTIONS P. LTD.

101,103, Vikas Surya Plaza, CU Block
L.S.C. Market, Pitam Pura, New Delhi-110 034
Ph : +91 11 27341616, 27341717, 27341718
E-mail: newindiapublishingagency@gmail.com
www: www.nipabooks.com

For customer assistance, please contact
Phone: + 91-11-27 34 17 17
Fax: + 91-11- 27 34 16 16
E-Mail: feedbacks@nipabooks.com

ISBN: 978-81-19002-81-8

Composed and Designed by NIPA®.

Preface

Is the life beyond the molecules? Our knowledge about the chemistry of life is not sufficient enough to answer this question. However, our quest to resolve the secret of life is continual. Today the biochemical wisdom is quite in depth about the building blocks of living beings i.e. Carbohydrates, Fates, Proteins and Nucleic acids. Enough understanding is also generated about the control, rate, extent of biological processes and multiplication of biological replicates. The blue print of life is there in the genomics and the life is throbbing everywhere due to physiological processes. The knowledge should be perpetuated down the generations to carry this lamp ahead on the path of wisdom. This is possible through teaching, learning and skill development. It is also required to test the knowledge gained . The mind accumulates, classify, analyze and synthesized logical facts to produce the analog for further inferences and hypotheses. After learning a huge text, the theories and practices are abstracted in the form of mind charts or brief summaries in the mind. After a long experience in the teaching and learning we have distillated those mind charts into "Biochemistry, Molecular Biology and Biotechnology - Instant Notes". The purpose of this collection is to quickly recall the understanding of Biochemistry, Genetics, Biotechnology up to post graduate level. This text will help to get command on the above subject for students appearing for JEE, JRF, SRF, NET, SET, ARS etc. and the teachers involved in couching these students.

Authors

Contents

CHAPTER - 1

Water, pH and Buffer

WATER IS VERY SPECIAL IN BIOLOGY

Most abundant chemical of life

Special properties:

Property	Biological Applications
Liquid over wide temperature range	Chemical reactions take place readily in liquids. Useful for circulatory systems. Many organisms live in a liquid matrix.
Excellent solvent	Good for chemical reactions, waste removal, delivery of food materials.
Ionizing solvent (high dielectric constant)	Ionizes salts, makes conductive solutions. Important for nerves and other excitable tissues.
Floats when it freezes	Ice floats on lakes and ponds. Less chance of freezing to bottom. Important for fish survival in winter.
Low viscosity (flows readily)	Important for circulatory systems. Less work for the heart.
High tensile strength	Allows trees to pull water hundreds of meters upward (transpiration) without breaking the column of water.
High surface tension (surface acts as though coated with tough film)	Allows many small animals to walk on water (water strider, basilisk lizard) or hang from the surface.
High heat capacity (large amount of heat required to raise water temperature)	Reduces daily temperature fluctuations. Climate is moderate near bodies of water (such as the bay). Useful for moving large amounts of heat in circulatory systems.
High heat of vaporization (large amount of heat removed when water evaporates)	Have a cooling effect on lakes and other bodies of water. Used in mammals for sweating (the only way the body can lose heat if the ambient temperature is above body temperature).
High heat conduction	Useful for removing the heat produced by biological reactions. Prevents overheating of the body.

These properties arise from the anatomy of the polar water molecule: a dipole with slight negative charge on the oxygen atom and slight positive charges on the hydrogen atoms- cause water molecules to stick together by hydrogen-bonding

Properties of Water

1) Polarity

Covalent bonds (electron pair is shared) between oxygen and hydrogen atoms with a bond angle of 104.5^o.

Oxygen atom is more electronegative that hydrogen atom → electrons spend more time around oxygen atom than hydrogen atom → result is a POLAR covalent bond.

Creates a permanent dipole in the molecule.

Can determine relative solubility of molecules "like dissolves like".

2) Hydrogen bonds

Due to polar covalent bonds → attraction of water molecules for each other.

Creates hydrogen bonds = attraction of one slightly positive hydrogen atom of one water molecule and one slightly negative oxygen atom of another water molecule.

The length of the bond is about twice that of a covalent bond.

Each water molecule can form hydrogen bonds with four other water molecules. It is weaker than covalent bonds (about 25x weaker).

Hydrogen bonds give water a **high melting point**.

Density of water decreases as it cools → water expands as it freezes → ice results from an open lattice of water molecules → less dense, but more ordered.

Hydrogen bonds contribute to water's **high specific heat** (amount of heat needed to raise the temperature of 1 gm of a substance 1^oC) - due to the fact that hydrogen bonds must be broken to increase the kinetic energy (motion of molecules) and temperature of a substance → temperature fluctuation is minimal.

Water has a **high heat of vaporization** - large amount of heat is needed to evaporate water because hydrogen bonds must be broken to change water from liquid to gaseous state.

3) Universal solvent

Water can interact with and dissolve other polar compounds and those that ionize (electrolytes) because they are hydrophilic.

Do so by aligning themselves around the electrolytes to form **solvation spheres** - shell of water molecules around each ion.

Solubility of organic molecules in water depends on polarity and the ability to form hydrogen bonds with water.

Functional groups on molecules that confer solubility:

carboxylates

protonated amines

amino

hydroxyl

carbonyl

As the number of polar groups increases in a molecule, so does its solubility in water.

4) Hydrophobic interactions

Nonpolar molecules are not soluble in water because water molecules interact with each other rather than nonpolar molecules → nonpolar molecules are excluded and associate with each other (known as the hydrophobic effect).

Nonpolar molecules are hydrophobic.

Molecules such as detergents or surfactants are **amphipathic** (have both hydrophilic and hydrophobic portions to the molecule).

Usually have a hydrophobic chain of 12 carbon atoms plus an ionic or polar end.

Soaps are alkali metal salts of long chain fatty acids - type of detergent.

e.g. sodium palmitate

e.g. sodium dodecyl sulfate (synthetic detergent)

All form micelles (spheres in which hydrophilic heads are hydrated and hydrophobic tails face inward.

Contain 80-100 detergent molecules.

Used to trap grease and oils inside to remove them.

5) Other noncovalent interactions in biomolecules

There are **four major noncovalent forces** involved in the structure and function of biomolecules:

A) hydrogen bonds

More important when they occur between and within molecules → stabilize structures such as proteins and nucleic acids.

B) hydrophobic interactions

Very weak.

Important in protein shape and membrane structure.

C) charge-charge interactions or electrostatic interactions (ionic bonds)

Occur between two oppositely charged particles.

Strongest noncovalent force that occurs over greater distances.

Can be weakened significantly by water molecules (can interfere with bonding).

D) van der Waals forces

Occurs between neutral atoms.

Can be attractive or repulsive, depending upon the distance of the two atoms.

Much weaker than hydrogen bonds.

The actual distance between atoms is the distance at which maximal attraction occurs.

Distances vary depending upon individual atoms.

6) Nucleophilic nature of water

Chemicals that are electron-rich (**nucleophiles**) seek electron-deficient chemicals (**electrophiles**).

Nucleophiles are negatively charged or have unshared pairs of electrons → attack electrophiles during substitution or addition reactions.

Examples of nucleophiles: oxygen, nitrogen, sulfur, carbon, water (weak).

Important in condensation reactions, where hydrolysis reactions are favored.

e.g. protein → amino acids

In the cell, these reactions actually only occur in the presence of hydrolases.

Condensation reactions usually use ATP and exclude water to make the reactions more favorable.

7) Ionization of water

Pure water ionizes slightly can act as an acid (proton donor) or base (proton acceptor).

$2H_2O \rightarrow H_3O^+ + OH^-$, but usually written

$H_2O \rightarrow H^+ + OH^-$

Equilibrium constant for water:

$$Keq = \frac{[H^+][OH^-]}{[H_2O]} = 1.8 \times 10^{-16} M \text{ at } 25°C$$

if $[H_20]$ is 55.5 M → 1 liter of H_2O is 1000 g

1 mole of H_2O is 18 g

Can rearrange equation to the following:

$1.8 \times 10^{-16}M(55.5 M) = [H+][OH-]$

$1.0 \times 10^{-14}M^2 = [H+][OH-]$

At equilibrium, $[H^+] = [OH^-]$, so

$1.0 \times 10^{-14}M^2 = [H+]^2$

$1.0 \times 10^{-7} = [H^+]$

8- pH scale

pH = - log [H+], so at equilibrium

$pH = -\log (1.0 \times 10^{-7})$

$= 7$

pH <7 is acidic, pH > 7 is basic or alkaline

1 change in pH units equals a 10-fold change in $[H^+]$

Acid Dissociation Constants of Weak Acids

A strong acid or base is one that completely dissociates in water.

e.g. $HCl \rightarrow H^+ + Cl^-$

A weak acid or base is one that does not; some proportion of the acid or base is dissociated, but the rest is intact.

A weak acid or base can be described by the following equation:

weak acid (H) → H^+ + A^- conjugate acid-base pair HA

proton donor conjugate base (conjugate acid)

Each acid has a characteristic tendency to lose its proton in solution.

The stronger the acid, the greater the tendency to lose that proton.

The equilibrium constant for this reaction is defined as the acid dissociation constant or Ka.

$$K_a = \frac{[H^+][\text{conjugate base or } A^-]}{[HA]}$$

$pK_a = -\log K_a$ similar to pH

The pK_a is a measure of acid strength. The more strongly dissociated the acid, the lower the pKa, the stronger the acid.

Hence,

$$K_a = \frac{[H^+][A^-]}{[HA]}$$

$$\log K_a = \log \frac{[H^+][A^-]}{[HA]}$$

$$\log K_a = \log [H^+] + \log \frac{[A^-]}{[HA]}$$

$$-\log[H^+] = -\log Ka + \log \frac{[A^-]}{[HA]}$$ Henderson-Hasselbach equation

H-H equation defined the pH of a solution in terms of pK_a and log of conjugate base and weak acid concentrations.

Therefore, if [A-] = [HA], then

$$pH = pK_a + \log 1$$

$$pH = pK_a$$

The pKa values of weak acids are determined by titration. Can calculate the pH of a solution as increasing amounts of base are added.

e.g. acetic acid titration curve

$$CH_3COOH \xrightarrow{OH^-} CH_3COO^- + H_2O$$

This is the sum of two reactions that are occurring:

$$H_2O \rightarrow H+ \ + OH-$$

$$CH_3COOH \rightarrow CH_3COO- \ + H^+$$

When add OH^- to solution, will combine with free $H^+ \rightarrow H_2O$ (pH rises as $[H^+]$ falls).

When this happens, CH_3COOH immediately dissociates to satisfy its equilibrium constant (law of mass action).

As add more OH-, increase ionization of CH_3COOH.

At the midpoint, 1/2 of CH_3COOH has been ionized and $[CH_3COOH] = [CH_3COO^-]$.

As you continue to add more OH^-, have a greater amount of ionized form compared to weak acid.

Finally reach a point where all the weak acid has been ionized.

This titration is completely reversible.

This titration curve shows that a weak acid and its anion can act as a buffer at or around the pK_a.

Important in cells where pH is critical.

Can also use this principle to determine whether amino acids are charged or not at different pHs or just physiological pH.

Can use the H-H equation to calculate pH of a solution knowing the information in Table 2.4 (pK_a values) and the ratios of the second term (don't need to know actual concentrations, just ratio).

If $[A^-] > [HA]$, then the pH of the solution is greater than pK_a of the acid.

If $[A^-] < [HA]$, then the pH of the solution is less than the pK_a of the acid.

Buffers

Solutions that prevent sudden changes in pH when bases or acids are added.

Consist of a weak acid and its conjugate base.

Work best at + 1 pH unit from $pK_a \rightarrow$ maximal buffering capacity.

Excellent example:

blood plasma-carbon dioxide- carbonic acid- bicarbonate buffer system

$$CO_2 + H_2O \rightarrow H_2CO_3 \rightarrow HCO_3^- + H^+$$

If $[H^+]$ increases (pH falls), momentary increase in $[H_2CO_3]$, and equation goes to the left.

Excess CO_2 is expired (increased respiration) to re-establish equilibrium.

Occurs in hypovolemia, diabetes, and cardiac arrest.

If $[H^+]$ falls (pH increases), H_2CO_3 will dissociate to release bicarbonate ion and hydrogen ion. This results in a fall in CO_2 levels in the blood. As a result, breathing slows.

Occurs in vomiting, hyperventilation (coming at equation from left).

Three Types of Chemical Bonds are Important in Biology

Chemical bonds are all due to electrical attractions

Type of Bond	Characteristics	Biological Importance
Covalent	Bonding electrons shared between 2 atoms.	This type of bond holds together the long chains of macromolecules. These molecules do not split apart in water.
Ionic	Complete transfer of electron from one atom to another. Oppositely charged atoms attract one another.	Compounds with ionic bonds split into ions in water. Ions conduct electricity. Gives specialized cells (nerve, muscle) excitable properties.
Hydrogen	Weaker than covalent or ionic bonds. Formed between a hydrogen covalently bonded to O or N and a second O or N. The second O or N may be on the same molecule or on another nearby molecule.	Water: makes water molecules stick together. Responsible for many of the strange properties of water. Proteins: cause protein chains to spiral and bend, giving unique shapes. DNA: hold together the 2 chains to form the double helix. Allow chains to "unzip" for replication and transcription.

About 30 Chemical Elements are Used by Living Cells

The most important elements in biology are:

C	Carbon	H	Hydrogen	O	Oxygen
N	Nitrogen	P	Phosphorus	S	Sulfur
Na	Sodium	K	Potassium	Mg	Magnesium
Ca	Calcium	Cl	Chlorine	Fe	Iron
I	Iodine	Mn	Manganese	Zn	Zinc

Roughly speaking, the top 6 elements (C, H, O, N, P and S) are involved in organic compounds (carbon compounds), while the others are involved in inorganic compounds. Carbon compounds tend to be covalent (see next paragraph) while inorganic compounds tend to be ionic. There are many exceptions to these generalizations, but they are useful nevertheless.

97.6% of your body weight is made up of C, H, O, N, P and S.

Oxidation and Reduction Reactions are Used in Cell Metabolism

Oxidation is the loss of an electron; reduction is the gain of an electron

Electron need not be completely removed from an atom- if it is shifted away from the atom in a covalent bond the atom is considered to be oxidized

Oxidation and reduction always occur together

Many biological reactions, especially those involved in photosynthesis (chloroplasts) and the Krebs cycle (mitochondria) involve electron transfers from one molecule to another

Cells are Very Sensitive to pH

H^+ and OH^- ions get special attention because they are very reactive

Substance which donates H^+ ions to solution = acid

Substance which donates OH^- ions to solution = base

pH = - log $[H^+]$ (logarithmic scale); $\{H^+\}$ must be in moles/liter

	Approximate pH	Common Examples
Strong Acids	0-2	Stomach acid (HCl), battery acid (H_2SO_4)
Weak Acids	3-6	Lemon juice, vinegar, rainwater
Neutral	7	Pure water
Weak Bases	8-11	Bicarbonate solution
Strong Bases	12-14	Solutions of lye (NaOH), oven cleaner (KOH)

Human blood pH is 7.4

Blood pH above 7.4 = alkalosis

Blood pH below 7.4 = acidosis

Human body must get rid of ~15 moles potential acid/day (mostly CO_2)

In neutralization H+ and OH- react to form water

pH changes change charges on molecules, especially proteins

pH is stabilized by buffers

CHAPTER - 2

Macromolecules

Cells Contain 4 Major Types of Giant Molecules (Macromolecules, Polymers)

Carbohydrates

Lipids

Proteins

Nucleic Acids

BIOLOGICAL POLYMERS ARE MADE FROM ABOUT 60 SMALL BUILDING BLOCKS

Cells contain thousands of types of giant molecules (macromolecules, polymers)

Polymers are macromolecules made from a single building block molecule

Most of the macromolecules are made from about 60 types of small building block molecules

Most macromolecules are linear: building blocks linked head to tail- a few have branches (some polysaccharides)

This is what you would get if you heated cells in hydrochloric acid overnight at 110 degrees C:

Building Blocks	Number of Kinds	Polymers
Amino Acids	20	Proteins
Fatty Acids	10	Storage Lipids/Membranes
Sugars & Relatives	10	Polysaccharides/Nucleic Acids
Nucleotides	5	Nucleic Acids
Others	15	All of Above

MAKING MACROMOLECULES INVOLVES DEHYDRATION AND REQUIRES ENERGY

Macromolecules naturally break down into their building blocks

Hooking the building blocks together to make polymers requires energy from ATP

ATP is the "energy currency of the cell" (a little like the Rs.)

Each time a bond is formed to make macromolecule and water is removed (dehydration)

When macromolecules break down spontaneously they take up water (hydrolysis); this is what happens in digestion

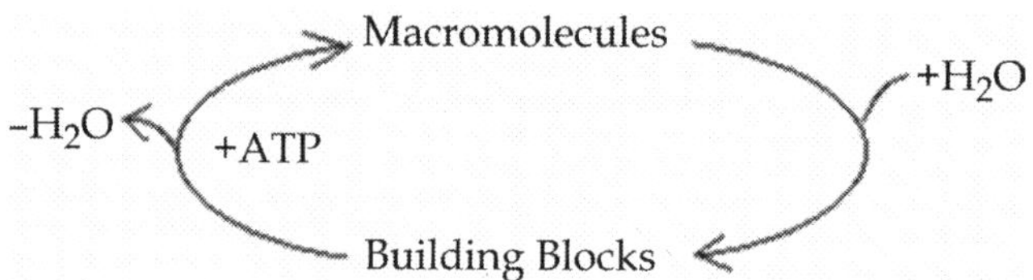

A. MACROMOLECULES - CARBOHYDRATES

Polysaccharides are Made from Small Sugar Molecules

Carbohydrates are small sugars and sugar polymers (polysaccharides)

Most of the sugars have 3-7 carbons; 6 is the most common

Hydrate refers to water.

Notice that most of the carbons in carbohydrates have OH (hydroxyl) on one side and H on the other. If you put the OH and H together you will get water. If you take common cane sugar (sucrose) and treat it with concentrated sulfuric acid it will pull the water off and you will see the black carbon left behind.

Each small sugar molecule contains either an aldehyde group or a keto group (the 3 below all have aldehyde groups at one end- to see keto groups look at the formulas for fructose or ribulose in your textbook)

These are 3 biological carbohydrates that are good to know. Glucose is the major source of energy in our bodies (and in the bodies of most organisms). It has 6 carbons and in the process of glycolysis is split into two 3 carbon pyruvic acids. If the cell is aerobic it is eventually oxidized in the Krebs cycle to CO_2 and water.

The other 2 sugars have 5 carbons. They are found in the molecules of heredity, DNA (deoxyribose) and RNA (ribose). Notice that deoxy ribose is missing the oxygen at the second position.

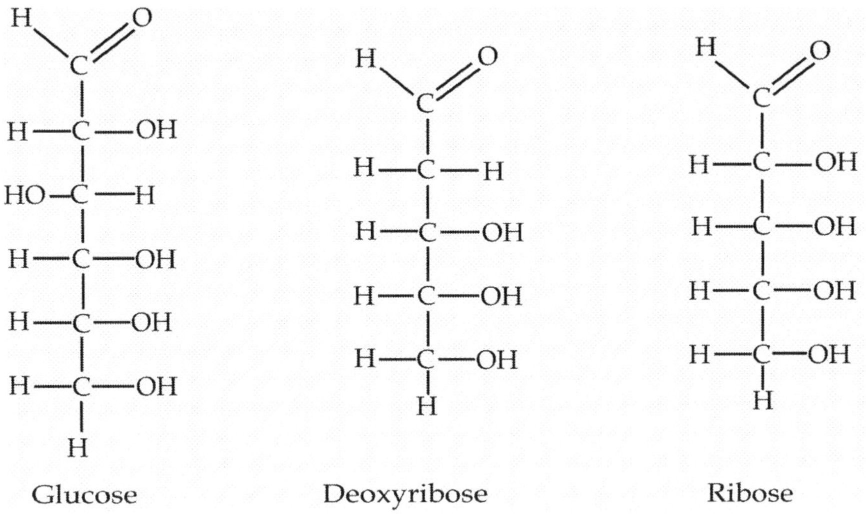

H—C=O
H—C—OH
HO—C—H
H—C—OH
H—C—OH
H—C—OH
H

Glucose

H—C=O
H—C—H
H—C—OH
H—C—OH
H—C—OH
H

Deoxyribose

H—C=O
H—C—OH
H—C—OH
H—C—OH
H—C—OH
H

Ribose

These 3 sugars are monosaccharides (1 sugar). Two sugars attached together = disaccharide (example = cane sugar or sucrose)

Sugars usually curl up into rings in water solutions. Here are the ring forms of deoxyribose and ribose:

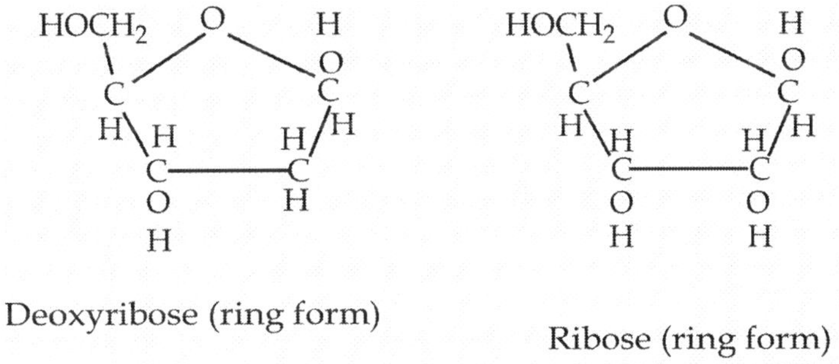

$HOCH_2$ O H O C C H H H H C C O H H

Deoxyribose (ring form)

$HOCH_2$ O H O C C H H H H C C O O H H

Ribose (ring form)

Polysaccharides are Used for Structure and Energy Storage

Glycogen: energy storage in animals (4 Calories/gram)

Starches: energy storage in plants

Cellulose: forms plant cell walls

Chitin: forms exoskeleton (external skeleton) of arthropods- contains amono group

Some sugars are attached to membrane proteins (example: ABO blood groups)

Different Polysaccharides Have Different Bonds Between the Sugars

In polysaccharides the sugars are linked at different positions; also the links may have different orientations

Both starch and cellulose are polymers of glucose, with the number 1 carbon of one sugar attached to the number 4 carbon of the next sugar.

Orientation of the bonds is different → quite different properties

We can use starch for food but not cellulose: we do not have digestive enzymes to split cellulose

Carbohydrates also called saccharides; most abundant molecules on earth.

Most are produced by photosynthesis.

Uses: Yield energy (ATP) to drive metabolic processes.

- Energy-storage molecules (i.e. glycogen, starch).
- Structural - cell walls and exoskeletons of some organisms.
- Carbohydrate derivatives found in coenzymes (FAD) and nucleic acids.

Can be described by the number of monomers they contain:

1) **monosaccharides** - $(CH_2O)n$ where n = 3-6; one sugar molecule
2) **oligosaccharides** - polymers from 2-20
3) **polysaccharides** - polymers of greater than 20 sugar residues
4) **glycoconjugates** - derivatives; attached to proteins, lipids, peptide chains

Monosaccharides

Also known as polyhydroxy aldehydes or ketones.

Classified based upon type of carbonyl group (-C-) and number of carbon atoms

aldose - sugar with aldehyde group

ketose - sugar with ketone group

There are two important trioses:

glyceraldehyde (aldotriose) and dihydroxyacetone (ketotriose)

Can have stereoisomers, but the D-form predominates.

Naming of sugars: xylose aldose

Xyulose ketose add "ul" before "ose".

Both aldoses and ketoses engage in intramolecular cyclization

Alcohol + aldehyde = **hemiacetal**

Alcohol + ketone = **hemiketal**

Both 5 and 6 carbon monosaccharides can form hemiacetals.

5-membered = **furanose**

6-membered = **pyranose**

Most oxidized carbon (C-1; attached to 2 oxygen atoms) is known as an **anomeric** carbon.

Can adopt either of two configurations.

- OH group down

- OH group up

Aldoses and ketoses equilibrate between cyclic and open forms.

Ring structures can adopt different conformations

5-member - twist (2 carbons approximately coplanar)

envelope (4 carbons approximately coplanar)

6-member - chair (more stable due to lack of steric hindrance from 6′ carbon) boat

Derivatives of monosaccharides

1) Sugar phosphates

   Metabolized as phosphate esters

2) Deoxy sugars

   Hydrogen atoms replaces -OH group on C-2.

   Important to structure of nucleic acids.

3) Amino sugars

   Amino group (NH-) substituted for -OH group in monosaccharide.

4) Sugar alcohols

   Replace carbonyl oxygen to form polyhydroxy alcohols

      e.g. glycerol → glyceraldehyde

   Replace "-ose" with "-itol".

   Ribose —> ribitol

5) Sugar acids

   Oxidation of carbonyl carbon or highest carbon.

glucose —> gluconate or glucuronate

Important in many polysaccharides.

6) Ascorbic acid

Derived from D-glucuronate.

Primates cannot do the conversion, so must be supplied in the diet.

Derived Sugars / Sugar Derivatives

Type of Derived sugars	No. of carbon atoms	Name of derived sugars	Occurance
Deoxysugar	5	2-Deoxy ribose	DNA
	6	L-Rhamnose	Component of cell wall
Aminosugar	6	D-Glucosamine	A major component of polysaccharide found in insects and crustaceans (chitin)
Polyol	6	Sorbitol	Berries
	6	Mannitol	Commercially prepared from mannose and fructose
Aldonic acid	6	Gluconic acid	-
Uronic acid	6	Glucuronic acid	Constituent of chondroitin sulfate
	6	Galacturonic acid	Constituent of pectin
Aldaric acid (Saccharic acid)	6	Glucaric acid	Oxidation product of glucose
	6	Mucic acid	Oxidation product of galactose

Disaccharides

Two monosaccharides joined by covalent bond called a **glycosidic linkage** via a condensation reaction.

Bond is created between the C-1 of one sugar and the -OH of another carbon

Examples:

1) **Maltose** - 2 glucose molecules joined by α-glycosidic bond

C-1 of one residue and C-4 of second residue

also known as α-D-glucopyranosyl-(1→ 4)-α-D-glucopyranose

2) **Cellobiose** - 2 glucoses joined by β-glycosidic bond; plant polysaccharide
3) **Lactose** - galactose and glucose in β-glycosidic bond; major carbohydrate in milk
4) **Sucrose** - glucose and fructose in $\alpha1 \rightarrow \beta2$ linkage; table sugar

Reducing and nonreducing sugars

- Some monosaccharides and most disaccharides have a reactive carbonyl group or anomeric carbon that can be oxidized.
- Examples: glucose, maltose, cellobiose, lactose
- Detected by the ability to reduce $Cu^{2+} \rightarrow Cu^{+}$ with Benedict's zreagent (blue —> red-orange).
- Non reducing sugars have both anomeric carbons in a glycosidic bond (e.g. sucrose).

Composition, sources and properties of common disccharides

Disaccharides	Constituent mono- saccharides	Linkage	Source	Properties
Reducing disaccharides				
Maltose	α-D-glucose + α-D-glucose	α(1→4)	Germinating cereal and malt	Forms osazone with phenylhydrazine. Fermentable by enzyme maltase present in yeast. Hydrolysed to two molecules of D-glucose. Undergoes mutarotation.
Lactose	β-D-galactose + β-D-glucose	β(1→4)	Milk. In trace amounts it can be seen in urine during pregnancy	It shows reactions of reducing sugars including mutarota-tion. Decomposed by alkali. Not fermentable by yeast. Hydrolysed to one molecule of galactose and one molecule of glucose by acids and the enzyme lactase

Contd...

Non reducing disaccharides				
Sucrose	a-D-glucose + b-D-fructose	a,b(1®2)	Sugar beet, sugar cane, sorghum and carrot roots	Fermentable. Hydrolysed by dilute acids or enzyme invertase (sucrase) to one molecule of glucose and one molecule of fructose. Relatively stable to reaction with alkali
Trehalose	α-D-glucose + α-D-glucose	α,α(1→1)	Fungi and yeast. It is stored as a reserve food supply in insect's hemolymph	It is hydrolysable by acids to glucose with difficulty. Not hydrolysed by enzymes.

Polysaccharides

Divided in two two classes:

1) **Homoglycans** - homopolysaccharides composed on one monosaccharide
2) **Heteroglycans**- heteropolysaccharides made of more than one type of monosaccharide

Often classified according to their biological role:

1) **Starch and Glycogen** - storage polysaccharides
    - Both are homoglycans.
    - Starch is storage form in plants and fungi.
    - Glycogen is storage form in animals.
    - Bacteria contain both.

Starch - mixture of **amylose** and **amylopectin**

- amylose is an unbranched polymer of 100-1000 D-glucose in an α-(1 → 4) glycosidic linkage.
- amylopectin is a branched polymer α (1→ 6) branches of residues in an α -(1 → 4) linkage; overall between 300-6000 glucose residues, with branches once every 25 residues; side chains are 15-25 residues long
    - **α -amylase** is an endoglycosidase found in human saliva but also plants that randomly hydrolyzes the α (1→4) bond of amylose and amylopectin.

- **β-amylase** is an exoglycosidase found in higher plants that hydrolyzes maltose residues from non-reducing ends of amylopectin.

Glycogen - branched polymer of glucose residues with branches every 8-12 residues with branches containing as many as 50,000 glucose residues

2) **Cellulose and Chitin** - structural polysaccharides

cellulose - straight chain homoglycan of glucose with β 1—> 4 linkages with alternating glucose molecules; ranges in size from 300-15,000 glucose residues

Extensive H-bonding within and between cellulose chains.

Makes bundles or fibrils → rigid.

chitin - linear polymer of N-acetylglucosamine residues

Alternating 180° with β - (1 → 4) linkage.

Lots of H-bonding between adjacent strands.

Glycoconjugates

Heteroglycans of three types:

1) **Proteoglycans**
   - Complexes of polysaccharides called glycosaminoglycans and core proteins.
   - Found in extracellular matrix of connective tissues.
   - Glycosaminoglycans are unbranched heteroglycans of disaccharide units (amino sugar, D-galactosamine or D-glucosamine, and alduronic acid). e.g. hyaluronic acid Found in cartilage and synovial fluid.
   - Proteoglycan cartilage

2) **Peptidoglycans**
   - Found in cell wall of bacteria.
   - Composed of alternating residues of N-acetylglucosamine and N-acetylmuramic acid joined by α- (1→ 4) linkages

3) **Glycoproteins**
   - Proteins with oligosaccharides attached.
   - Carbohydrate chains are from 1-30 residues in length.

- Examples: enzymes, hormones, structural proteins, transport proteins.
- Found in eucaryotic cells.
- Can be attached to proteins with one of two configurations:
  1) O-linked - carbohydrate bonded to -OH of serine or threonine
  2) N-linked - carbohydrate (usually N-acetylglucosamine) linked to asparagines

Typical Comparison of storage and structural polysachharides

Storage polysachharides	Structural polysachharides
· Occurs in the storage organs such as seeds and tubers.	· Important constituents of plant cell wall.
· Include both homo and heteropolysaccharides	• Include both homo and heteropolysaccharides
· Starch and inulin are homopolysaccharides made from glucose and fructose, respectively	· Cellulose is a homopolysaccharide made up of glucose
· Arabinogalactan is an example for storage heteropolysaccharide	· Hemicelluloses are heterpolysaccharides containing pentoses, hexoses and monosaccharide derivatives
· The configuration of major linkages are a-type	· The configuration of major linkages are β-type

Chemical Properties of Carbohydrates

- Reactions of monosaccharides are due to the presence of hydroxyl (-OH) and the potentially free aldehyde (-CHO) or keto (>C=O) groups.
- Sugars in weak alkaline solutions undergo isomerization to form 1,2-enediol followed by the formation of a mixture of sugars.
- Under strong alkaline conditions sugar undergo caramelization reactions.
- Sugars are classified as either reducing or non-reducing depending upon the presence of potentially free aldehyde or keto groups. The reducing property is mainly due to the ability of these sugars to reduce

metal ions such as copper or silver to form insoluble cuprous oxide, under alkaline condition. The aldehyde group of aldoses is oxidized to carboxylic acid. This reducing property is the basis for qualitative (Fehling's, Benedict's, Barfoed's and Nylander's tests) and quantitative reactions. All monosaccharides are reducing. In the case of oligosaccharides, if the molecule possesses a free aldehyde or ketone group it belongs to reducing sugar (maltose and lactose). If the reducing groups are involved in the formation of glycosodic linkage, the sugar belongs to the non- reducing group (trehalose, sucrose, raffinose and stachyose).

- When reducing sugars are heated with phenylhydrazine at pH 4.7 a yellow precipitate is obtained. The precipitated compound is called as osazone. One molecule of reducing sugar reacts with three molecules of phenylhydrazine.
- D-mannose and D-fructose form same type of osazone as that of D-glucose since the configuration of C-3, C-4, C-5 and C-6 is same for all the three sugars. The osazone of D-galactose is different. Different sugars form osazone at different rates. For example, D-fructose forms osazone more readily than D-glucose. The osazones are crystalline solids with characteristic shapes, decomposition points and specific optical rotations. The time of formation and crystalline shape of osazone is utilized for identification of sugars. If methyl phenylhydrazine is used instead of phenylhydrazine in the preparation of osazone, only ketoses react. This reaction serves to distinguish between aldose and ketose sugars.
- The hydroxyl group formed as a result of hemiacetal formation in monosaccharides react with methanol and HCl to form methyl α- and β-glycosides. The derivaties of each sugar are named according to the name of the sugar, that is, the derivaties of glucose as glucosides, of galactose as galactosides and of arabinose as arabinosides etc. Glycosides are acid-labile but are relatively stable at alkaline pH. Since the formation of glycosides convert the aldehydic group to an acetal group, the glucosides are not a reducing sugars.
- Glycosides are also formed with a non-sugar component, the aglycone. The sugars which are connected to the non-sugar moiety, are pentoses, hexoses, branched sugars or deoxy or dideoxy sugars. The chain length varies from one to five monosaccharide sugar residues per glycosides. Apart from O-glycosides, three other classes of glycosides are found in higher plants namely S-glycosides, N-glycosides and C-glycosides.

- Monosaccharides are generally stable to hot dilute mineral acids though ketoses are appreciably decomposed by prolonged action.
- Heating a solution of hexoses in a strong non-oxidising acidic conditions, hydroxy methyl furfural is formed. The hydroxymethyl furfural from hexose is usually oxidized further to other products When phenolic compounds such as resorcinol, α-naphthol or anthrone are added, mixture of coloured compounds are formed .
- The molisch test used for detecting carbohydrate in solution is based on following principle. When conc. H_2SO_4 is added slowly to a carbohydrate solution containing α-naphthol, a pink color is produced at the juncture. The heat generated during the reaction hydrolyse and dehydrate it to produce furfural or hydroxymethyl furfural which then react with α-naphthol to produce the pink color.
- When sugars are treated with appropriate acid anhydride or acid chloride under proper conditions, the hydroxyl groups get esterified and form sugar esters

Roles of Carbohydrates in Biology

Carbohydrates serve as information-rich molecules that guide many biological processes.

Examples include:

1) Asialoglycoprotein receptor
   - Present in liver cells; binds to asialoglycoproteins to remove them from circulation
   - Presence of sialoglycoprotein prevents glycoproteins such as antibodies and peptide hormones from being internalized
   - Presence of sialic acid on terminal galactose on these proteins mark the passage of time; when they are removed (usually by the protein itself), the glycoproteins are removed from circulation
2) Lectins
   - Carbohydrate-binding proteins of plant origin.
   - Contain 2 or more binding sites for carbohydrate units → cross-link or agglutinate erythrocytes and other cells.
3) Many viruses and bacteria can gain entry into host cells via carbohydrates displayed on cell surface.
   - Influenza virus contains a hemagglutinin protein that recognizes sialic acid residues on cells lining respiratory tract.

- *Neisseria gonorrhoeae* infects human genital or oral epithelial cells because of recognition of cell surface carbohydrates; other cells lack these carbohydrates.

4) Interaction of sperm with ovulated eggs
   - Ovulated eggs contain zona pellucida, an extracellular coat made of O-linked oligosaccharides.
   - Sperm cells have receptor for these carbohydrates.
   - Binding of sperm to egg causes release of proteases and hyaluronidase, which dissolve zona pellucida to allow sperm entry.

5) Selectins
   - Carbohydrate-binding adhesion proteins that mediate binding of neutrophils and other leukocytes to sites of injury in the inflammatory response.

6) Homing receptor of lymphocytes
   - Homing is phenomenon in which lymphocytes tend to migrate to lymphoid sites from which they were originally derived.
   - Mediated by carbohydrates on lymphocyte surface and endothelial lining of lymph nodes.

B. MACROMOLECULES - LIPIDS

Lipids are esters of glycerol with fatty acids

Lipids are water-insoluble that are either hydrophobic (nonpolar) or amphipathic (polar and nonpolar regions).

Classification of Lipids

Simple lipids

- Esters of fatty acids with glycerol and monohydric alcohols
- Depending upon the constituent alcohols they are further subdivided into fats or oils and waxes
- Fats, also termed as triacylglycerols are esters of fatty acids with glycerol e.g. Plants-vegetable oils; Animals-ghee and butter
- Waxes are esters of fatty acids and alcohols other than glycerol e.g., Plant wax-carnauba wax; Insect wax-beeswax; Animal wax -lanolin.

Compound lipids

- Esters containing chemical groups in addition to alcohol and fatty acids
- Depending upon the chemical groups they are further subdivided into phospholipids, glycolipids, sulpholipids and lipoproteins
- Phospholipids contain phosphate group. Phopholipids are further grouped as glycerophospholipids, if the constituting alcohol is glycerol (eg. Lecithin) or as sphingophospholipids, if the alcohol is sphingosine (eg. Sphingomyelin).
- Glycolipids contain hexose units preferably galactose alongwith fatty acids and alcohol eg. cerebrosides
- Plant sulpholipids contain sulfated hexose with fatty acids and alcohol
- Lipoproteins contain protein subunits along with lipids. Depending upon density and lipid compound they are further classified as VLDL. LDL and HDL.

Derived lipids

- Substances derived form simple and compound lipids by hydrolysis of alcohols, fatty acids, aldehydes, ketones, sterols and hydrocarbons

Lipids are water-insoluble that are either hydrophobic (nonpolar) or amphipathic (polar and nonpolar regions). Lipids are in many ways the most diverse of the biological macromolecules, since they are something of a rag-tag bunch of leftovers. Lipids are pretty much everything in the cell that isn't very water soluble, and chemically they don't have a great deal in common with one another. The best known lipids are probably the fatty acids, so that is where we shall start.

THE FATTY ACIDS

The fatty acids are long chain carboxylic acids synthesised by the condensation and reduction of acetyl coenzyme-A units by fatty acid synthase. The more important ones have nonsystematic names in wide use. Stearic and palmitic acids are saturated (no double bonds), oleic acid is monounsaturated, and linoleic and linolenic are polyunsaturated. All these common fatty acids are cis (E) fatty acids. Because of the kinks in the chain caused by the double bonds, the unsaturated fatty acids tend to be liquids at room temperature (they are less easy to pack together to form a solid). Bacteria and plants (which cannot thermoregulate) will use more unsaturated acids in their cell membranes when they are exposed to cold: this helps to maintain membrane fluidity.

Molecular structure

Name of Fatty acid

Lauric acid: saturated C_{12}

Myristic acid: saturated C_{14}

Palmitic acid: saturated C_{16}

Stearic acid : saturated C_{18}

Oleic acid : monounsaturated C_{18}

Linoleic acid : diunsaturated C_{18}

γ -Linolenic acid : triunsaturated C_{18}

Arachidonic acid : tetraunsaturated C_{20}

The nomenclature of these acids is rather complicated. There are at least five systems in use. Here are some of the above in the different systems. The delta system numbers the double bonds from the carboxyl group (the w carbon), whereas the omega system indicates where the first double bond is counting from the other end of the molecule (the w carbon).

Trivial	Systematic	Colon	Delta	Omega
Stearic acid	Octadecanoic acid	18:0	Octadecanoic acid	-
Palmitic acid	Hexadecanoic acid	16:0	Hexadecanoic acid	-
Oleic acid	E-Octadec-9-enoic acid	18:1n9	cis-Δ^9-octadecenoic acid	ω9
Linoleic acid	9E, 12E-Octadeca-9, 12-dienoic acid	18:2n9	cis, cis-$\Delta^{9,12}$-octadecadienoic acid	ω6
Linolenic acid	6E, 9E, 12E-Octadeca-6, 9, 12-trienoic acid	18:3n6	cis, cis, cis -$\Delta^{6,9,12}$-octadecatrienoic acid	ω3

Essential Fatty Acids

Human body is unable to synthesise all fatty acids found in the body. Those fatty acids that are not synthesised in the body but required for normal body growth and maintenance are called as essential fatty acids. These fatty acids are to be supplied through diet. Linoleic and linolenic acids are essential fatty acids. The longer chain fatty acids can be synthesised by the body from dietary linoleic and α linolenic acids. Arachidonic acid is essential but it can be synthesised by the body from linolenic acid. It is also present in the meat. Linoleic acid is grouped under n 6 family because the 6th carbon from methyl end possesses the double bond.

Properties pf Fat

Physical

- They are insoluble in water and soluble in organic solvents.
- Pure triacylglycerols are tasteless, odourless, colourless and neutral in reaction.
- They have lesser specific gravity (density) than water and therefore float in water.
- Though fats are insoluble in water, they can be broken down into minute droplets and dispersed in water. This is called emulsification.
- They contain hydrophilic colloidal particles such as proteins, carbohydrates and phospholipids which act as stabilizing agents.
- Emulsification greatly increases the surface area of the fat and this is an essential requisite for digestion of fat in the intestine.

Chemical

- Fat is hydrolysed to yield three molecules of fatty acid and one molecule of glycerol. The hydrolysis of fat is effected by alkali and enzyme.
- The process of alkali hydrolysis is called 'saponification'.
- The alkali salt of fatty acid resulting from saponification is soap
- Hydrolysis of triacylglycerol may be accomplished enzymatically through the action of lipases.
- Development of disagreeable odour and taste in fat or oil upon storage is called rancidity. Rancidity reactions may be due to hydrolysis of

ester bonds (hydrolytic rancidity) or due to oxidation of unsaturated fatty acids (oxidative rancidity).

- The partial hydrolysis of the triacylglycerol to mono and diacylglycerol is called Hydrolytic rancidity. The hydrolysis is hastened by the presence of moisture, warmth and lipases present in fats or air. In fats like butter which contains a high percentage of volatile fatty acids, hydrolytic rancidity produces disagreeable odour and taste due to the liberation of the volatile butyric acid. Butter becomes rancid more easily in summer.
- The unsaturated fatty acids are oxidised at the double bonds to form peroxides, which then decompose to form aldehydes and acids of objectionable odour and taste. (Oxidative rancidity)

Constants of Fats and Oils

Since fats and oils form essential nutrient of human diet, it is necessary to identify a pure fat or to determine the proportion of different types of fat or oil mixed as adulterant in edible oils and fats like butter and ghee. With an adequate knowledge of the characteristic composition of fats or oils, it is possible to identify the fat or oil under investigation. The chemical constants also give an idea about the nature of fatty acids present in fats or oils. Eventhough gas chromatographic method is available to identify and quantify the fatty acids present in fat or oil, the physical and chemical constants are still used in routine public health laboratories where such sophisticated facilities are lacking.

Physical constants

- Specific gravity : Since different oils have different specific gravity, any variation from normal value shows mixture of oils.
- Refractive index : Fats have definite angles of refraction. Variation from the normal value indicates adulteration of fats or oils.
- Solidification point or setting point : Solidification point is the temperature at which the fat after being melted, sets back to solid or just solidifies. Each fat has a specific solidification point.

Chemical constants

- Saponification number : It is defined as milligrams of KOH required to saponify 1 gm of fat or oil. Saponification number is high for fat or oil containing low molecular weight or short chain fatty acids and vice versa. It gives a clue about the molecular weight and size of the fatty acid in the fat or oil.

- Iodine Number : It is defined as the number of grams of iodine taken up by 100 grams of fat or oil. Iodine number is a measure of the degree of unsaturation of the fatty acid. Since the quantity of the iodine absorbed by the fat or oil can be measured accurately, it is possible to calculate the relative unsaturation of fats or oil.
- Reichert Meisel number(R.M.number) : It is defined as the number of millilitres of 0.1 N alkali required to neutralise the soluble volatile fatty aicds contained in 5 gm of fat. It measures the volatile soluble fatty acids. It is confined to butter and coconut oil. The determination of Reichert Meisel number is important to the food chemist because it helps to detect the adulteration in butter and ghee. Reichert Meisel value is reduced when animal fat is used as adulterant in butter or ghee.
- Polanski number : It is defined as the number of millilitres of 0.1 N potassium hydroxide solution required to neutralise the insoluble fatty acids (not volatile with steam distillation) obtained from 5 gm of fat.Ghee may be adulterated by the addition of insoluble, non volatile fatty acids (by addition of animal fat). This can be tested by finding out the Polanski number
- Acetyl number : It is defined as the amount in millilitres of potassium hydroxide solution required to neutralise the acetic acid obtained by saponification of 1 gm of fat or oil after acetylation. Some fatty acids contain hydroxyl groups. In order to determine the proportion of these, they are acetylated by means of acetic anhydride. This results in the introduction of acetyl groups in the place of free hydroxyl groups. The acetic acid in combination with fat can be determined by titration of the liberated acetic acid from acetylated fat or oil with standard alkali. Acetyl number is thus a measure of the number of hydroxyl groups present in fat or oil.
- Acid number : It is defined as the milligram of potassium hydroxide required to neutralise the free fatty acids present in one gram of fat or oil. Acid number indicates the amount of free fatty acids present in fat or oil. The free fatty acid content increases with age of the fat or oil.

There are Many Types of Lipids

1) Fatty acids

- The simplest with structural formula of R-COOH where R = hydrocarbon chain.

- They differ from each other by the length of the tail, degree of unsaturation, and position of double bonds.
- pKa of -COOH is 4.5-5.0 → ionized at physiological pH.
- If there is no double bond, the fatty acid is saturated.
- If there is at least one double bond, the fatty acid is unsaturated.
- Monounsaturated fatty acids contain 1 double bond; polyunsaturated fatty acids have >2 double bonds.
- IUPAC nomenclature $=a^n$ represents where double bond occurs as you count from the carboxyl end.

e.g.	-enoate	one double bond
	-dienoate	2 "
	-trienoate	3 "
	-tetraenoate	4 "

- Can also use a colon separating 2 numbers, where the first number represents the number of carbon atoms and the second number indicates the location of the double bonds.

  e.g. linoleate 18:2 $\alpha^{9,12}$ or cis,cis -$\alpha^{9,12}$ octadecadienoate
- Physical properties differ between saturated and unsaturated fatty acids.

  Saturated = solid at RT; often animal source; e.g. lard

  Unsaturated = liquid at RT; plant source; e.g. vegetable oil
- The length of the hydrocarbon tails influences the melting point.
- As the length of tails increases, melting points increases due to number of van der Waals interactions.
- Also affecting the melting point is the degree of unsaturation.
- As the degree of unsaturation increases, fatty acids become more fluid → melting point decreases (kinks in tails decrease number of van der Waals interactions).
- Fatty acids are also an important sources of energy.

  9 kcal/g vs. 4 kcal/g for carbohydrates and proteins.

2) Triacylglycerols

- Also called triglycerides.

- Made of 3 fatty acyl residues esterified to glycerol.
- Very hydrophobic, neutral in charge → can be stored in anhydrous form.
- Long chain, saturated triacylglycerols are solid at RT (fats).
- Shorter chain, unsaturated triacylglycerols are liquid at RT (oils).
- Lipids in our diet are usually ingested as triacylglycerols and broken down by lipases to release fatty acids from their glycerol backbones
- Also occurs in the presence of detergents called bile salts.
  - Form micelles around fatty acids that allow them to be absorbed by intestinal epithelial cells.
  - Transported through the body as lipoproteins.

```
   H
   |    O   H H H H H HH H H HH H H H
H-C-O--C-C-C-C-C-C-C-C-C-C-C-C-C-C-C-H
   |        H H H H H H HH H H HH H HH
   |    O   H H H H H HH H H HH H H H
H-C-O--C-C-C-C-C-C-C-C-C-C-C-C-C-C-C-H
   |        H H H H H H HH H H HH H HH
   |    O   H H H H H HH H H HH H H H
H-C-O--C-C-C-C-C-C-C-C-C-C-C-C-C-C-C-H
   |        H H H H H H HH H H HH H HH
   H
```

Triglyceride (Neutral Fat)

3) Glycerophospholipids

- Main components of cell membranes.
- Are amphipathic and form bilayers.
- All use glycerol 3-phosphate as backbone.
- Simplest is phosphatidate = 2 fatty acyl groups esterified to glycerol 3-phosphate.
- Often, phosphate is esterified to another alcohol to form...
  - phosphatidylethanolamine
  - phosphatidylserine
  - phosphatidylcholine

- Enzymes called phospholipases break down biological membranes.
    - A-1 = hydrolysis of ester bond at C-1.
    - A-2 = hydrolysis of ester bond at C-2; found in pancreatic juice.
    - C = hydrolysis of P-O bond between glycerol and phosphate to create phosphatidate.
    - D = same

4) Sphingolipids

- Second most important membrane constituent.
- Very abundant in mammalian CNS.
- Backbone is sphingosine (unbranched 18 carbon alcohol with 1 trans C=C between C-4 and C-5), NH_3^+ group at C-2, hydroxyl groups at C-1 and C-3.
- Ceramides are intermediates of sphingolipid synthesis.
- There are three families of sphingolipids:
    1) sphingomyelin - phosphocholine attached to C-1 hydroxyl group of ceramide; present in the myelin sheaths around some peripheral nerves.
    2) cerebrosides - glycosphingolipid; has 1 monosaccharide (galactose) attached by α-glycosidic linkage to C-1 of ceramide; most common is galactocerebroside, which is abundant in nervous tissue.
    3) gangliosides - glycosphingolipid containing N-acetylneuraminic acid; present on all cell surfaces.
- Hydrocarbon tails embedded in membrane with oligosaccharides facing extracellularly.
- Probably used as cell surface markers, e.g. ABO blood group antigens.
- Inherited defects in ganglioside metabolism $\rightarrow$ diseases, such as Tay-Sachs disease.

5) Steroids

Steroids have 4 rings, which makes the molecule very flat and stiff. The best known steroid is cholesterol. Too much cholesterol causes clogging of the arteries, but in the right amount it is very useful. It is essential for cell membranes (it keeps them from falling apart) and is a precursor for several hormones, including cortisone, aldosterone and both male and female sex hormones.

- Called isoprenoids because their structure is similar to isoprene.
- Have 4 fused rings: 3 6-membered rings (A,B,C) and 1 5-membered ring (D).
- Cholesterol is an important component of cell membranes of animals, but rare in plants and absent in procaryotes.
- Also have mammalian steroid hormones (estrogen, androgens) and bile salts.
- Differ in length of side chain at C-17, number and location of methyl groups, double bonds, etc.
- Cholesterol's role in membranes is to broaden the phase transition of cell membranes ---> increases membrane fluidity because cholesterol disrupts packing of fatty acyl chains.

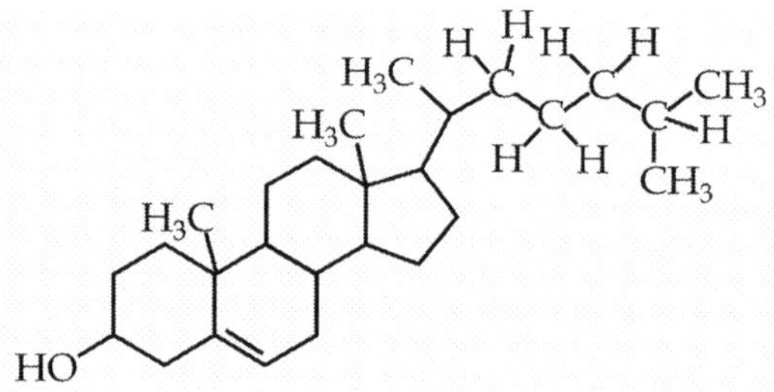

Cholesterol

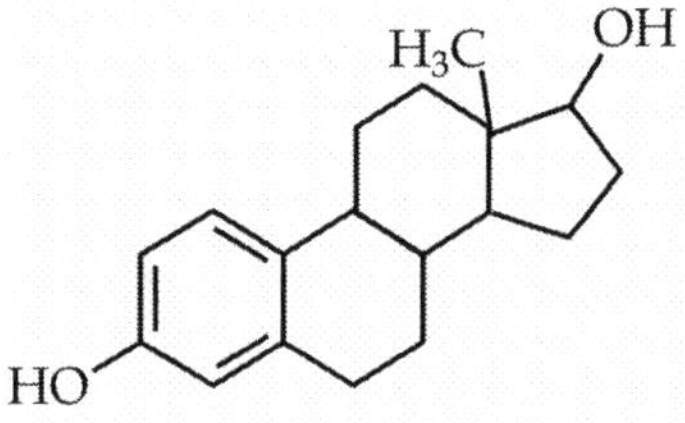

Estradiol
(female sex hormone)

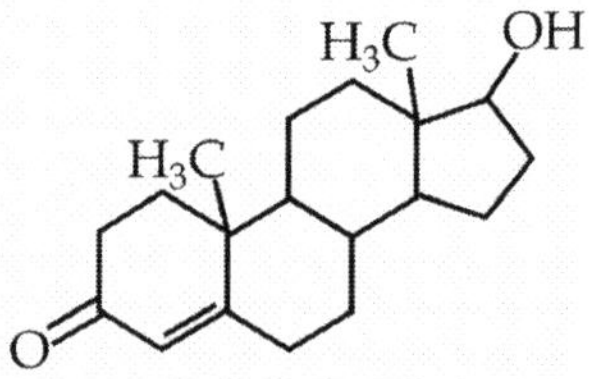

Testosterone
(male sex hormone)

6) Other lipids not found in membranes

- **waxes** - nonpolar esters of long chain fatty acids and alcohols

  very water insoluble

  high melting point → solid at outdoor/RT.

  Roles: protective coatings of leaves, fruits, fur, feathers, exoskeletons.

- **eicosanoids** - 20 carbon polyunsaturated fatty acids

  e.g. prostaglandins - affect smooth muscle → cause constriction; bronchial constriction of asthmatics; uterine contraction during labor
- **limonene** - smell of lemons
- **bactoprenol** - involved in cell wall synthesis
- **juvenile hormone I** - larval development of insects

C. MACROMOLECULES - PROTEINS

Functions of Proteins:

- catalysts - enzymes for metabolic pathways
- storage and transport - e.g. myoglobin and hemoglobin
- structural - e.g. actin, myosin
- mechanical work - movement of flagella and cilia, microtubule movement during mitosis, muscle contraction
- decoding information - translation and gene expression
- hormones and hormone receptors
- specialized functions - e.g. antibodies

Proteins are Made up of Exactly 20 Types of Amino Acids

Chemistry of amino acids

Amino acids are the building blocks of proteins. They all have the basic structure shown below:

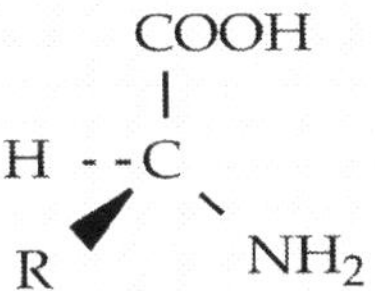

L-amino acid
CO-R-N goes anticlockwise

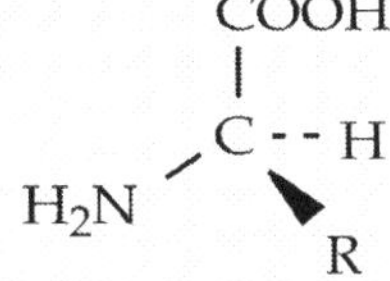

D-amino acid
CO-R-N goes clockwise

The proteins in our bodies are made of 20 different L-amino acids strung together like beads on a string.

All 20 have a part in common: a central carbon attached to a carboxyl, an amino and a hydrogen atom.

The differences between the 20 types are in the side groups that can be designated as R.

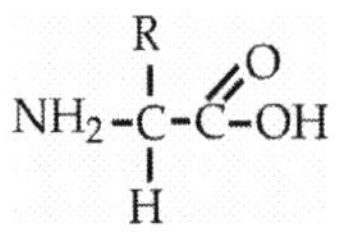

```
       R
       I  //O
NH2 -C-C-OH
       I
       H
```

Note the amino (NH2), carboxyl (COOH) and hydrogen groups (H) which are found on all amino acids. The differences in amino acids are in the 20 different types of R groups.

Basic Amino Acid Structure

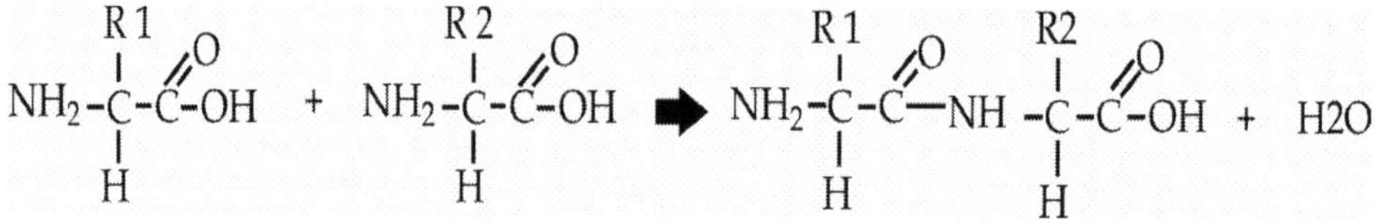

```
      R1                  R2                      R1          R2
       I  //O              I  //O                  I  //O      I  //O
NH2-C-C-OH  +  NH2-C-C-OH  ➡  NH2-C-C—NH -C-C-OH  +  H2O
       I                   I                       I           I
       H                   H                       H           H
```

Formation of a Peptide Bond

Two amino acids can combine, with the carboxyl of one attaching to the amino group of the other. Water is removed. This process also requires energy in the form of ATP.

Amino acids are grouped based upon the properties and structures of side chains.

1) Aliphatic (R groups consist of carbons and hydrogens)
   - glycine - R=H smallest a.a. with no chiral center
   - alanine - R=CH3 methyl group
   - valine R = branched; hydrophobic; important in protein folding
   - leucine R= 4 carbon branched side chain
   - isoleucine R = 2 chiral centers
   - proline R = ring; puts bends or kinks in proteins; contains a secondary amino group
2) Aromatic (R groups have phenyl ring)
   - phenylalanine - very hydrophobic
   - tyrosine - hydrophobic, but not as much because of polar groups
   - tryptophan - "

   Absorb UV light at 280 nm ® used to estimate protein
3) Sulfur-containing R groups
   - methionine - sulfur is internal (hydrophobic)
   - cysteine - sulfur is terminal ® highly reactive; can form disulfide bonds

4) Side chains with alcohols
    - serine - α-hydroxyl groups → hydrophilic
    - threonine -
5) Basic R groups
    - histidine - hydrophilic side chains - + charged at neutral pH
    - lysine - "
    - arginine - strong base
6) Acidic R groups and amide derivatives
    - aspartate - • carboxyl group - confer - charges on proteins
    - glutamate - a carboxyl group
    - asparagine - amide of aspartate - side groups uncharged → polar
    - glutamine - amide of glutamate - "
    - Amide groups can form H bonds with atoms of other polar amino acids.

Classification of Amino Acids

The protein amino acids are classified according to the chemical nature of their R groups as aliphatic, aromatic, heterocyclic and sulphur containing amino acids. More meaningful classification of amino acids is based on the polarity of the R groups. The polarity of the R groups varies widely from totally non-polar to highly polar. The 20 amino acids are classified into four main classes.

Amino acids with non-polar or hydrophobic, aliphatic R groups

This group of amino acids includes glycine, alanine, valine, leucine, isoleucine and proline. The hydrocarbon R groups are non-polar and hydrophobic. The side chains of alanine, valine, leucine and isoleucine are important in promoting hydrophobic interactions within protein structures. On the other hand, the imino group of proline is held in a rigid conformation and reduces the structural flexibility of the protein.

Amino acids with non-polar aromatic R groups

This group includes phenylalanine, tyrosine and tryptophan. All these amino acids participate in hydrophobic interactions, which is stronger than aliphatic R groups because of stacking one another. Tyrosine and tryptophan are more polar than phenylalanine due to the presence of hydroxyl group in tyrosine and nitrogen in the indole ring of tryptophan. The absorption of ultraviolet (UV) light at 280 nm by tyrosine, tryptophan and to a lesser extent by phenylalanine is responsible for the characteristic strong absorbance of light by proteins. This property is exploited in the characterization and quantification of proteins.

Amino acids with polar, uncharged R groups

This group of amino acids includes serine, threonine, cysteine, methionine, asparagine and glutamine. The hydroxyl group of serine and threonine, the sulphur atom of cysteine and methionine and the amide group of asparagine and glutamine, contribute to the polarity. The R groups of these amino acids are more hydrophilic than the non-polar amino acids.

Amino acids with charged R groups

Acidic : The two amino acids with acidic R groups are aspartic and glutamic acids. These amino acids have a net negative charge at pH 7.0.

Basic : This group includes lysine, arginine and histidine. The R groups have a net positive charge at pH 7.0. The lysine has a second ε-amino group; arginine has a positively charged guanidino group; and histidine has an imidazole group.

Essential Amino Acids

Most of the prokaryotic and many eukaryotic organisms (plants) are capable of synthesizing all the amino acids present in the protein. But higher animals including man possess this ability only for certain amino acids. The other amino acids, which are needed for normal functioning of the body but cannot be synthesized from metabolic intermediates, are called essential amino acids. These must be obtained from the diet and a deficiency in any one of the amino acids prevents growth and may even cause death. Methionine, Threonine, Tryptophan, Valine, Isoleucine, Leucine, Phenylalanine and Lysine are the essential amino acids, however, Histidine and Arginine are considered semi essential amino acids as it can be partly synthesized by the body.

Properties of Amino Acids

Physical

- Amino acids are white crystalline substances. Most of them are soluble in water and insoluble in non-polar organic solvents (e.g., chloroform and ether).
- Aliphatic and aromatic amino acids particularly those having several carbon atoms have limited solubility in water but readily soluble in polar organic solvents.

- They have high melting points varying from 200-300°C or even more.
- They are tasteless, sweet or bitter. Some are having good flavour.
- Amphoteric nature of amino acids as they contain both acidic (COOH) and basic (NH_2) groups. The amino acids possessing both positive and negative charges are called zwitterions.
- Isomerism - All amino acids except proline, found in protein are *-amino acids because NH_2 group is attached to the *-carbon atom, which is next to the COOH group. Examination of the structure of amino acids reveals that except glycine, all other amino acids possess asymmetric carbon atom at the position.

Chemical properties

- Reaction with formaldehyde (Formal titration): An amino acid solution is treated with excess of neutralized formaldehyde solution, the amino group combines with formaldehyde forming dimethylol amino acid which is an amino acid formaldehyde complex Hence the amino group is protected and the proton released is titrated against alkali. This method is used to find out the amount of total free amino acids in plant samples.
- Reaction with nitrous acid: Nitrous acid reacts with the amino group of amino acids to form the corresponding hydroxyacids and liberate nitrogen gas.
- Reaction with ninhydrin: Ninhydrin is a strong oxidizing agent. When a solution of amino acid is boiled with ninhydrin, the amino acid is oxidatively deaminated to produce ammonia and a ketoacid. The keto acid is decarboxylated to produce an aldehyde with one carbon atom less than the parent amino acid. The net reaction is that ninhydrin oxidatively deaminates and decarboxylates *-amino acids to CO_2, NH_3 and an aldehyde. The reduced ninhydrin then reacts with the liberated ammonia and another molecule of intact ninhydrin to produce a purple coloured compound known as Ruhemann's purple.
- This ninhydrin reaction is employed in the quantitative determination of amino acids. Proteins and peptides that have free amino group(s) (in the side chain) will also react and give colour with ninhydrin.
- Decarboxylation: The carboxyl group of amino acids is decarboxylated to yield the corresponding amines. Thus, the vasoconstrictor agent, histamine is produced from histidine. Histamine stimulates the flow of gastric juice into the stomach and the dilation and constriction of

specific blood vessels. Excess reaction to histamine causes the symptoms of asthma and various allergic reactions.

Proteins are Amino Acids Linked Together by Peptide Bonds

- In proteins the carboxyl group (COOH) of one amino acid attaches to the amine group (NH2) of the next to form a peptide bond.
- The reaction requires ATP
- Water is removed
- Protein chains are not branched
- Note that the R groups stick out to the side of the protein chain
- Each protein has a terminal amino group on one end and a terminal carboxyl on the other end

Classification of Protein

Proteins are classified based on their

(A) Solubility and composition

(B) Function

(C) Size & Shape

(A) Classification of proteins based on solubility and composition

Proteins are again divided into three main groups as simple, conjugated and derived proteins.

(I) Simple proteins

Simple proteins yield on hydrolysis, only amino acids. These proteins are further classified based on their solubility in different solvents as well as their heat coagulability.

Albumins

Albumins are readily soluble in water, dilute acids and alkalies and coagulated by heat. Seed proteins contain albumin in lesser quantities. Albumins may be precipitated out from solution using high salt concentration, a process called 'salting out'. They are deficient in glycine. Example - Serum albumin and Ovalbumin (egg white).

Globulins

Globulins are insoluble or sparingly soluble in water, but their solubility is greatly increased by the addition of neutral salts such as sodium chloride. These proteins are coagulated by heat. They are deficient in methionine. Example - Serum globulin, Fibrinogen, Myosin of muscle and Globulins of pulses.

Prolamins

Prolamins are insoluble in water but soluble in 70-80% aqueous alcohol. Upon hydrolysis they yield much proline and amide nitrogen. They are deficient in lysine. Example - Gliadin of wheat and Zein of corn.

Glutelins

Glutelins are insoluble in water and absolute alcohol but soluble in dilute alkalies and acids. They are plant proteins. Example - Glutenin of wheat.

Histones

Histones are small and stable basic proteins and contain fairly large amounts of basic amino acid, histidine. They are soluble in water, but insoluble in ammonium hydroxide. They are not readily coagulated by heat. They occur in globin of hemoglobin and nucleoproteins.

Protamines

Protamines are the simplest of the proteins. They are soluble in water and are not coagulated by heat. They are basic in nature due to the presence of large quantities of arginine. Protamines are found in association with nucleic acid in the sperm cells of certain fish. Tyrosine and tryptophan are usually absent in protamines.

Albuminoids

These are characterized by great stability and insolubility in water and salt solutions. These are called albuminoids because they are essentially similar to albumin and globulins. They are highly resistant to proteolytic enzymes. They are fibrous in nature and form most of the supporting structures of animals. They occur as chief constituent of exoskeleton structure such as hair, horn and nails.

(II) Conjugated or compound proteins

These are simple proteins combined with some non-protein substances known as prosthetic groups. The nature of the non-protein or prosthetic groups is the basis for the sub classification of conjugated proteins.

Nucleoproteins

Nucleoproteins are simple basic proteins (protamines or histones) in salt combination with nucleic acids as the prosthetic group. They are the important constituents of nuclei and chromatin.

Mucoproteins

These proteins are composed of simple proteins in combination with carbohydrates like mucopolysaccharides, which include hyaluronic acid and chondroitin sulphates. On hydrolysis, mucopolysaccharides yield more than 4% of amino-sugars, hexosamine and uronic acid e.g., ovomucoid from egg white. Soluble mucoproteins are neither readily denatured by heat nor easily precipitated by common protein precipitants like trichloroacetic acid or picric acid. The term glycoproteins is restricted to those proteins that contain small amounts of carbohydrate usually less than 4% hexosamine.

Chromoproteins

These are proteins containing coloured prosthetic groups e.g., haemoglobin, flavoprotein and cytochrome.

Lipoproteins

These are proteins conjugated with lipids such as neutral fat, phospholipids and cholesterol.

Metalloproteins

These are metal-binding proteins. A *-globulin, termed transferrin is capable of combining with iron, copper and zinc. This protein constitutes approximately 3% of the total plasma protein. Another example is ceruloplasmin, which contains copper.

Phosphoproteins

These are proteins containing phosphoric acid. Phosphoric acid is linked to the hydroxyl group of certain amino acids like serine in the protein, Example- casein of milk.

(III) Derived proteins

These are proteins derived by partial to complete hydrolysis from the simple or conjugated proteins by the action of acids, alkalies or enzymes. They include two types of derivatives, primary-derived proteins and secondary-derived proteins.

Primary-derived proteins

These protein derivatives are formed by processes causing only slight changes in the protein molecule and its properties. There is little or no hydrolytic cleavage of peptide bonds.

Proteans

Proteans are insoluble products formed by the action of water, dilute acids and enzymes. These are particularly formed from globulins but are insoluble in dilute salt solutions. Example - myosan from myosin, fibrin from fibrinogen.

Metaproteins

These are formed by the action of acids and alkalies upon protein. They are insoluble in neutral solvents.

Coagulated proteins

Coagulated proteins are insoluble products formed by the action of heat or alcohol on natural proteins. Example - cooked meat and cooked albumin.

Secondary-derived proteins

These proteins are formed in the progressive hydrolytic cleavage of the peptide bonds of protein molecule. They are roughly grouped into proteoses, peptones and peptides according to average molecular weight. Proteoses are hydrolytic products of proteins, which are soluble in water and are not coagulated by heat. Peptones are hydrolytic products, which have simpler structure than proteoses. They are soluble in water and are not coagulated by heat. Peptides are composed of relatively few amino acids. They are water-soluble and not coagulated by heat. The complete hydrolytic decomposition of the natural protein molecule into amino acids generally progresses through successive stages as follows:

Protein $\rightarrow$ Protean $\rightarrow$ Metaprotein $\rightarrow$

Proteoses $\rightarrow$ Peptones $\rightarrow$ Peptides $\rightarrow$ amino acids

(B) Classification of proteins based on function

Proteins are classified based on their functions as -

(I) Catalytic proteins - Enzymes

The most striking characteristic feature of these proteins is their ability to function within the living cells as biocatalysts. These biocatalysts are called as enzymes. Enzymes represent the largest class. Nearly 2000 different kinds of enzymes are known, each catalyzing a different kind of reaction. They enhance the reaction rates a million fold.

(II) Regulatory proteins - Hormones

These are polypeptides and small proteins found in relatively lower concentrations in animal kingdom but play highly important regulatory role in maintaining order in complex metabolic reactions.Example- growth hormone, insulin etc.

(III) Protective proteins - Antibodies

Some proteins have protective defense function. These proteins combine with foreign protein and other substances and fight against certain diseases.Example - immunoglobulin. These proteins are produced in the spleen and lymphatic cells in response to foreign substances called antigen. The newly formed protein is called antibody which specifically combines with the antigen which triggered its synthesis thereby prevents the development of diseases. Fibrin present in the blood is also a protective protein.

(IV) Storage proteins

A major class of proteins which has the function of storing amino acids as nutrients and as building blocks for the growing tissues. Storage proteins are source of essential amino acids, which cannot be synthesized by human beings. The major storage protein in pulses is globulins and prolamins in cereals. But in rice the major storage protein is glutelins. Albumin of egg and casein of milk are also storage proteins.

(V) Transport proteins

Some proteins are capable of binding and transporting specific types of molecules through blood. Haemoglobin is a conjugated protein composed of colourless basic protein, the globin and ferroprotoporphyrin or haem. It has the capacity to bind with oxygen and transport through blood to various tissues. Myoglobin, a related protein, transports oxygen in muscle. Lipids bind to serum proteins, principally, albumin and transported as lipoproteins in the blood.

(VI) Toxic proteins

Some of the proteins are toxic in nature. Ricin present in castor bean is extremely toxic to higher animals in very small amounts. Enzyme inhibitors such as trypsin inhibitor bind to digestive enzyme and prevent the availability of the protein. Lectin, a toxic protein present commonly in legumes, agglutinates red blood cells. A bacterial toxin causes cholera, which is a protein. Snake venom is protein in nature.

(VII) Structural proteins

Some proteins serve as structural materials or as important components of extra cellular fluid. Examples of structural proteins are myosin of muscles, keratin of skin and hair and collagen of connective tissue. Carbohydrates, fats, minerals and other cellular components are organized around such structural proteins that form the molecular framework of living material.

(VIII) Contractile proteins

Proteins like actin and myosin function as essential elements in contractile system of skeletal muscle.

(IX) Secretary proteins

Fibroin is a protein secreted by spiders and silkworms to form webs and cocoons.

(X) Exotic proteins

Antarctic fishes live in -1.90C waters, well below the temperature at which their blood is expected to freeze. These fishes are prevented from freezing by antifreeze glycoproteins present in their body.

(C) Classification of proteins based on size and shape

Based on size and shape, the proteins are also subdivided into globular and fibrous proteins. Globular proteins are mostly water-soluble and fragile in nature. Example- enzymes, hormones and antibodies. Fibrous proteins are tough and water-insoluble. They are used to build a variety of materials that support and protect specific tissues. Example- Skin, Hair, Fingernails and Keratin.

Conformation of proteins

Conformation of a protein refers to the three-dimensional structure in its native state. There are many different possible conformations for a molecule as large as a protein. Four types of structural organization can be distinguished in proteins:

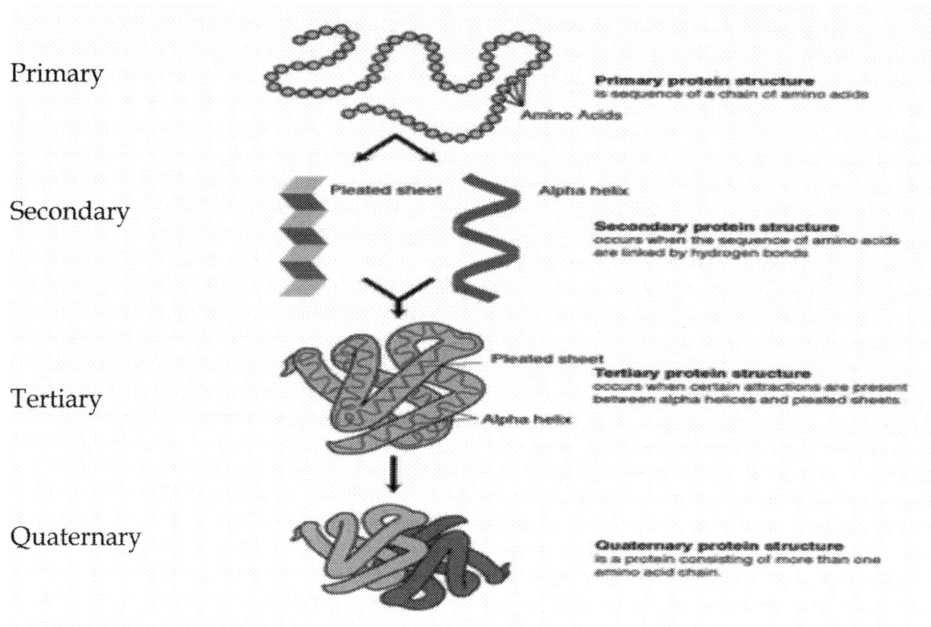

The Shapes of Proteins are Determined by Hydrogen Bonding and by the Numbers and Sequence of the Side Groups

- Primary structure of a protein is the sequence of amino acids. Each protein has a unique sequence determined by the genetic code (no substitutions allowed)
- Secondary structure of a protein is caused by regular hydrogen bonding between carbonyl and amino groups of the protein. This H-bonding causes the protein to coil and fold in various ways

  alpha helix: spiral structure common in fibrous proteins (hair)

  beta pleated sheat: H-bonding between adjacent parallel strands of protein (silk)
- Tertiary structure: caused by bonding between amino acid side groups.

  ionic (between carboxyls & amines)

  hydrophobic

  disulfide

  H-bonds
- Quaternary structure: produced by binding of 2 or more protein subunits together to form a functional unit

Example: hemoglobin if formed by association of 2 alpha and 2 beta chains

Protein Properties are Determined by their Shapes

- A typical protein will have 200 to 1000 amino acids. A molecule this big can bend into millions of different shapes, but only one is correct
- Proteins will not function unless folded into proper shape. Like any machine, the parts must be in the right positions before it can work
- Many proteins have"active sites" produced by side groups brought together by folding
- Protein shape is disrupted by many agents, especially heat- this is called denaturation
- Heat sterilizes by denaturing proteins of bacteria and viruses

Proteins Do Most of the Work of the Cell

- Every cell has thousands of different types of proteins, each specialized to do a certain job
- Some proteins are structural: control shape of cells and bind cells together

  Example: collagen- binds all of the cells of the body together
- Chemical reactions of the cell are controlled by protein enzymes
- Protein pumps move things across the cell membrane
- Proteins give mobility:

  Muscles

  Flagella

  "Molecular motors"
- Defend the body against foreign invaders: antibodies
- Receptors: required for signalling in endocrine and nervous systems

Protein Folding and Stability

- Once a polypeptide is made, it then folds into its characteristic three-dimensional shape.
- As the protein folds, initial interactions then initiate further interactions -called the cooperativity of folding.

- Folding occurs in less than a second.
- Protein folding and stabilization depend upon noncovalent forces, including the hydrophobic effect, hydrogen bonding, van der Waals interactions, and charge-charge interactions.
- Although individually weak, collectively they are strong.
- The weakness gives the protein flexibility to change conformations.
- Once in place, the collective effect keeps the protein in its proper shape.
- No actual protein folding pathway is known, however, the structure of some intermediates has been described.
- It appears that hydrophobic effects are very important initially, such that the protein "collapses" onto itself.
- Then some parts of secondary structure begin to form.
- Then motifs form, followed by the stable, completely folded protein.

The Hydrophobic Effect

- Proteins are more stable when their hydrophobic R-groups are in the interior of a protein and away from water.
- Nonpolar side chains then interact with each other.
- Polar side chains remain in contact with water on the protein surface.

Hydrogen Bonding

- Hydrogen bonds in a helices, b sheets and turns form first as a protein folds $\rightarrow$ defined regions of secondary structure.
- Many hydrogen bonds ultimately form between polypeptide backbone and water, between backbone and R-groups, between R-groups, and between R-groups and water.
- Those hydrogen bonds within interior of protein are more stable than those on the surface because these bonds do not then compete with water molecules.

Van der Waals Interactions and Charge-Charge Interactions

- Van der Waals contacts between nonpolar side chains are also important.
- Charge-charge interactions contribute minimally to protein stability because most ionic bonds are found on the surface of a protein.

Chaperones

- Protein folding does not involve a random search for the proper conformation.
- Secondly, the final shape of a protein is dependent upon its primary structure.
- Small proteins can fold properly in vitro, but larger ones need the help of molecular chaperones.
- Chaperones are proteins that assist with protein folding by binding to proteins before they are completely folded.
- They prevent the formation of incorrectly folded intermediates that may trap a polypeptide into an improper form.
- They also bind to protein subunits and prevent them from aggregating and precipitating before then are assembled into a multisubunit protein.
- Most chaperones are heat-shock proteins. Originally found when cells were subjected to temperature stress, which tends to make proteins denature.
- The major heat-shock protein is HSP-70, present in all eukaryotes and prokaryotes.
- Most highly conserved protein known → indicates the very important role of HSP-70 in folding.
- Chaperones usually bind to the hydrophobic portions of a protein and prevent them from interacting with water or at least coming into contact with water molecules.

How Protein Structure is Related to Function

Collagen

- Major component of connective tissue of vertebrates.
- Consists of three left-handed helical chains coiled around each other in a right-handed supercoil.
- Each helix has 3 amino acids per turn and a pitch of 0.94 nm → more extended than an α-helix.
- Stability of the collagen helix is achieved via interchain hydrogen bonds.
- Helical regions consist of the amino acids -Gly-X-Y, where X is usually proline and Y is usually hydroxyproline.

- For each -Gly-X-Y triplet, one hydrogen bond forms between the amide hydrogen atom of glycine in one chain and the carbonyl oxygen of an adjacent chain.
- There are no intrachain hydrogen bonds.
- Hydroxyproline and hydroxylysine are made from proline and lysine after the protein has been synthesized, i.e. an enzyme does the hydroxylation.
- In mammals, vitamin C is necessary for adequate hydroxylation.
- People who suffer from scurvy lack sufficient amounts of vitamin C in their diet.
- Develop skin lesions, fragile blood vessels (susceptible to bruising), loose teeth, and bleeding gums.
- Collagen triple helices are arranged in a staggered fashion to give rise to very strong fibers.
- There are some covalent cross-links between the side chains of some lysine and hydroxylysine residues to form Schiff bases between carbonyl groups and amines.

Myoglobin and Hemoglobin

Vertebrates must supply and deliver a constant amount of oxygen to tissues for aerobic respiration. This is done in two ways:

1) development of circulatory system that delivers oxygen to cells
2) use of oxygen-carrying molecules to overcome oxygen's low solubility in water. e.g. myoglobin and hemoglobin

The ability of myoglobin or hemoglobin to bind oxygen depends upon a heme group (prosthetic group).

Heme consists of an organic part (protoporphyrin) and iron atom.

Iron atom in center can form 6 bonds: 4 with nitrogens from protoporphyrin and 2 on either side of plane.

Iron atom can be in ferrous (+2) or ferric (+3) state → ferrohemoglobin and ferromyoglobin and ferrihemoglobin and ferrimyoglobin. Only +2 state can bind oxygen

Myoglobin structure determined with x-ray crystallography in mid 1950's.

Molecule has several important features:

1) extremely compact
2) 75% of structure in a-helix (8 helices, named A, B, C, ...H).

3) 4 of helices are terminated by proline residue
4) main-chain peptide groups are planar
5) little empty space inside molecule; interior consists almost entirely of nonpolar residues; amino acids that are amphipathic oriented so that hydrophilic portions face exterior; only polar amino acids in interior are 2 histidines, which are part of binding site.

Heme group located in crevice in myoglobin molecule.

Iron atom is bonded to histidine in F8 (histidine); the oxygen-binding site on iron is located on other side of heme plane (E7).

Binding of oxygen to heme must occur in a bent, end-on orientation.

If only a small portion of the protein binds oxygen, why have the rest of the protein?

Heme exposed to oxygen by itself rapidly oxidizes to +3, which cannot bind oxygen.

Heme is much less susceptible to oxidation because not only allows heme to bind oxygen, but it is a reversible process.

Carbon monoxide is a poison because it combines with ferromyoglobin and ferrohemoglobin to block oxygen transport.

CO's binding affinity is about 200x stronger than that for oxygen.

If allow CO to interact with isolated iron porphyrins, the iron, carbon, and oxygen atoms are in a linear array.

If allow CO to interact with myoglobin or hemoglobin, CO axis is bent, as in oxygen binding because of steric hinderance from His E7 $\rightarrow$ greatly weakens the interaction of CO with the heme.

Biological significance

CO is produced within cells in the breakdown of heme $\rightarrow$ about 1% of binding sites on hemoglobin and myoglobin are blocked by CO.

If affinity was close to that of isolated iron porphyrins $\rightarrow$ massive poisoning.

Bottom line: function of a prosthetic group is modulated by its polypeptide environment.

Hemoglobin consists of 4 polypeptide chains, 2 of one type, 2 of another, held together by noncovalent bonds.

Each polypeptide contains a heme group and oxygen binding site.

Embryos and fetuses have zeta chains and epsilon chains; zeta is replaced by alpha, epsilon chains are replaced with gamma, then beta chains.

The three-dimensional structures of myoglobin and hemoglobins are very similar → myoglobin resembles a chains of hemoglobin.

Odd because amino acid sequence is not very similar → different amino acid sequences can specify similar 3-D structures.

Those amino acids found to be invariant (do not change) are those directly bonded to heme iron or hold helices together.

The nonpolar character of interior of molecule is conserved → important in binding heme group and stabilizing 3-D structure of each subunit.

Hemoglobin is more intricate than myoglobin.

1) transports protons, carbon dioxide, and oxygen
2) is an allosteric protein
3) binding of oxygen to hemoglobin is cooperative
4) affinity of hemoglobin for oxygen is pH dependent; same true for CO_2
5) hemoglobin also regulated by 2,3-bisphosphoglycerate (BPG)

If look at oxygen dissociation curves for myoglobin and hemoglobin, find many differences:

1) saturation of myoglobin is higher at all oxygen pressures than hemoglobin → myoglobin has higher affinity for oxygen than does hemoglobin P50 for myoglobin is 1 torr; P50 for hemoglobin is 26 torr
2) oxygen dissociation curve of myoglobin is hyperbolic; that of hemoglobin is sigmoidal → binding of oxygen to hemoglobin is cooperative (seen in Hill plot)

Biological significance of cooperativity

Enables hemoglobin to deliver nearly twice as much oxygen under typical physiological conditions as it would if binding sites were independent

Physical and Chemical Properties of Proteins

Physical properties

- Pure proteins are generally tasteless, though the predominant taste of protein hydrolysates is bitter.
- Pure proteins are odourless.
- Because of the large size of the molecules, proteins exhibit many properties that are colloidal in nature.

- Proteins, like amino acids, are amphoteric and contain both acidic and basic groups.
- They possess electrically charged groups and hence migrate in an electric field.
- Many proteins are labile and readily modified by alterations in pH, UV radiation, heat and by many organic solvents.
- The absorption spectrum of protein is maximum at 280 nm due to the presence of tyrosine and tryptophan, which are the strongest chromophores in that region. Hence the absorbance of the protein at this wavelength is adapted for its determination.

Denaturation of protein - The comparatively weak forces responsible for maintaining secondary, tertiary and quaternary structure of proteins are readily disrupted with resulting loss of biological activity. This disruption of native structure is termed denaturation.

Physical and chemical factors are involved in the denaturation of protein

- Heat and UV radiation supply kinetic energy to protein molecules causing their atoms to vibrate rapidly, thus disrupting the relatively weak hydrogen bonds and salt linkages. This results in denaturation of protein leading to coagulation. Enzymes easily digest denatured or coagulated proteins.
- Organic solvents such as ethyl alcohol and acetone are capable of forming intermolecular hydrogen bonds with protein disrupting the intramolecular hydrogen bonding. This causes precipitation of protein.
- Acidic and basic reagents cause changes in pH, which alter the charges present on the side chain of protein disrupting the salt linkages.
- Salts of heavy metal ions (Hg^{2+}, Pb^{2+}) form very strong bonds with carboxylate anions of aspartate and glutamate thus disturbing the salt linkages. This property makes some of the heavy metal salts suitable for use as antiseptics.

Renaturation - Renaturation refers to the attainment of an original, regular three-dimensional functional protein after its denaturation.

When active pancreatic ribonuclease A is treated with 8M urea or *-mercaptoethanol, it is converted to an inactive, denatured molecule. When urea or mercaptoethanol is removed, it attains its native (active) conformation.

Chemical properties

- Colour reactions of proteins - The colour reactions of proteins are of importance in the qualitative detection and quantitative estimation of proteins and their constituent amino acids. Biuret test is extensively used as a test to detect proteins in biological materials.
- Biuret reaction - A compound, which is having more than one peptide bond when treated with Biuret reagent, produces a violet colour. This is due to the formation of coordination complex between four nitrogen atoms of two polypeptide chains and one copper atom.
- Xanthoproteic reaction - Addition of concentrated nitric acid to protein produces yellow colour on heating, the colour changes to orange when the solution is made alkaline. The yellow stains upon the skin caused by nitric acid are the result of this xanthoproteic reaction. This is due to the nitration of the phenyl rings of aromatic amino acids.
- Hopkins-Cole reaction - Indole ring of tryptophan reacts with glacial acetic acid in the presence of concentrated sulphuric acid and forms a purple coloured product. Glacial acetic acid reacts with concentrated sulphuric acid and forms glyoxalic acid, which in turn reacts with indole ring of tryptophan in the presence of sulphuric acid forming a purple coloured product.

D. MACROMOLECULES - NUCLEIC ACIDS

Nucleic Acids are the Molecules of Heredity

Two major types: DNA & RNA

Both types have code which specifies the sequence of amino acids in proteins

DNA = archival copy of genetic code, kept in nucleus, protected

RNA = working copy of code, used to translate a specific gene into a protein, goes into cytoplasm & to ribosomes, rapidly broken down

RNA is a molecule with a single strand

DNA is a double strand (a double helix; helix = spiral) held together by hydrogen bonds between the bases (A always bonds to T; C always bonds to G)

The Building Blocks for Nucleic Acids are Nitrogen Bases, Sugars and Phosphate

DNA and RNA are made of nucleotides strung together like beads

Nucleotides also have other functions in energy transfer and as signal molecules.

Nucleotides are made from 3 components: a nitrogen base (purine or pyrimidine), a 5 carbon sugar (ribose or deoxyribose) and 3 phosphates.

Nitrogen Base **Purine or Pyrimidine**	**Nucleoside** **Base + Sugar**	**Nucleotide** **Base + Sugar + 3 Phosphates**
Adenine (purine)	Adenosine	Adenosine triphosphate
Guanine (purine)	Guanosine	Guanosine triphosphate
Cytosine (pyrimidine)	Cytidine	Cytidine triphosphate
Thymine (pyrimidine)	Thymidine	Thymidine triphosphate
Uracil (pyrimidine)	Uridine	Uridine triphosphate

There are 5 important nitrogen bases. They are usually given a one letter code: adenine = A; guanine = G; cytosine = C; thymine = T; uracil = U.

The bases A, C, G and T are found in DNA.

In RNA the bases are A, C, G and U, that is T is replaced by U.

These bases, in groups of 3, are the basis of the genetic code; sugar and phosphate make up the backbone

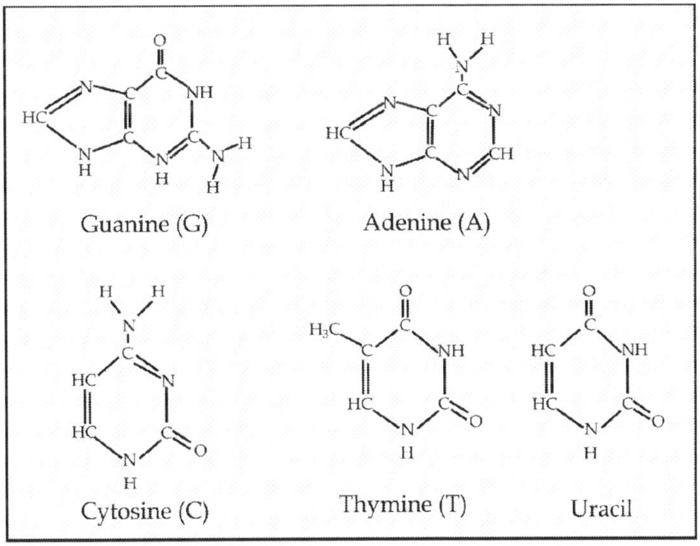

Guanine (G) Adenine (A)

Cytosine (C) Thymine (T) Uracil

The nitrogen bases are ring compounds with nitrogen in the rings.

Those with 2 rings are called purines, while those with a single ring are known as pyrimidines.

In DNA the 2 strands of the helix are held together by hydrogen bonds between the A and T, and between the C and G bases.

A nucleotide looks like this:

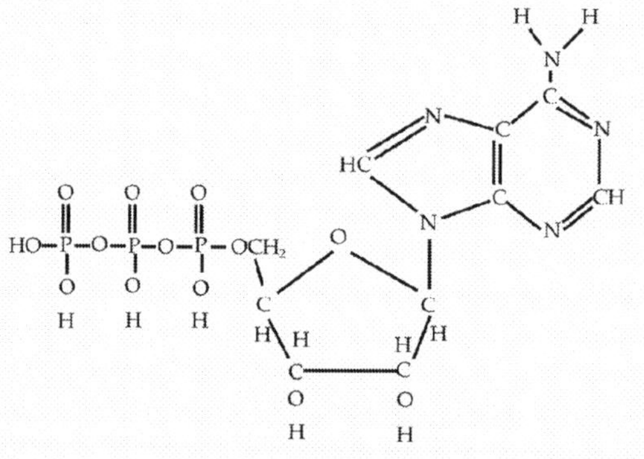

Adenosine Triphosphate (ATP: the energy currency of the cell)

ATP is used in almost all energy transactions of the cell

Useful energy is stored in the 4 bonds between the phosphate groups

Summary of Biological Macromolecules

Macromolecule	Building Blocks	Functions
Polysaccharides	Sugars	Energy storage (4 Cal/gm) Structure (cell walls, exoskeletons)
Lipids: triglycerides	Fatty acids, glycerol	Energy storage (9 Cal/gm)
Lipids: phospholipids	Fatty acids, glycerol, phosphate, polar groups	Cell membranes
Proteins	Amino acids: 20 types	Cell structure Enzymes Molecular motors (muscle, etc) Membrane pumps & channels Hormones & receptors Immune system: antibodies
Nucleic Acids: DNA	4 Bases: A, C, G, T Deoxyribose sugar, phosphate	Storage of hereditary information (genetic code)
Nucleic Acids: RNA 3 types: m-RNA t-RNA r-RNA	4 Bases: A, C, G, U Ribose sugar, phosphate	Protein synthesis: m-RNA: working copy of genetic code for a gene t-RNA & r-RNA: translation of the code

Chapter - 3

Energy in Biology

ALL MATTER IS IN MOTION

Everything moves:

In the atom electrons rotate around the nucleus

In gases, liquids and even solids molecules move away from other molecules

An accumulation of matter (i.e., a baseball) can move as a unit

The average velocity of movement is the distance travelled divided by the time

Bodies in motion have energy; the faster the velocity the more energy

Forces Cause Bodies to Accelerate and Pull Molecules Apart

Forces are pushes and pulls

Some forces require contact between 2 objects; others act at a distance (force travels through space)- gravity, electrical attraction

Forces applied to a body will cause its velocity to change (accelerate)

Acceleration = change in velocity/time

The larger the force the greater the acceleration

A body moving at a constant velocity (not accelerating) has balanced forces acting on it

In chemistry forces can pull bonds apart and make new ones

Pressure = force/area

Energy is the Capacity to do Work

Work = energy = force X distance

Some of the types of work done by cells:

- Building macromolecules
- Cell division
- Pumping ions and other materials across cell membranes
- Moving: flagella, ameboid movement, muscles, etc.
- Making light- bioluminescence
- Heating the cell

Some energy units: erg, joule, foot-pound, calorie, BTU, therm

Energy Can be Transformed from One Type to Another

Energy exists in different forms:

- Kinetic: energy of motion
- Chemical: bonds between atoms
- Light: vibratory energy, travels through space at immense velocity (300,000 kilometers/sec = 186,000 miles/sec)
- Sound: vibrations through matter. Much slower than light, cannot travel through a vacuum
- Heat: causes movement of molecules
- Electrical:

One type of energy can be converted into another

Some biological transformations of energy

- Photosynthesis: converts light from sun into chemical bond energy
- Muscle contraction: converts chemical energy from ATP into energy of movement
- Ion pumps: some convert chemcal energy from ATP into electrical energy (ion gradients)
- Brown fat in hibernating animals: converts chemical energy into heat

Energy is Conserved: Cannot be Created or Destroyed

This principle is called the First Law of Thermodynamics

Efficiency of energy conversion is never 100%- some of the energy is always converted to heat

Example:Usually in plant photosynthesis 95% of the light absorbed by the plant is converted to heat- only 5% is converted to chemical bond energy. Even if the optimum light wavelength is used the conversion of light to bond energy is only 27%.

Power is the Rate of Doing Work

Power = rate of energy use = energy/time

Units: watts, Calories/day, horsepower, METs, ATPs/sec

Energy Can be Stored in Chemical Bonds and Ion Gradients

Energy has 2 basic manifestations: kinetic and potential

Kinetic energy is the energy of motion

Potential energy is latent or stored- it can be tapped to produce motion or do work

Two examples of mechanical storage of energy:

Water in a lake or reservoir at high altitude: when it runs downhill the stored energy can be used to generate electricity, etc.

Energy stored in a stretched spring: can be released to make spring contract forcefully- when we run we store some energy in stretched tendons

In biology energy is often stored in chemical bonds of macromolecules:

Starch: major carbohydrate storage molecule of plants: 4 Cal/gm

Glycogen: major carbohydrate storage molecule of animals: 4 Cal/gm

Triglycerides: lipid storage molecule used by both plants and animals: 9 Cal/gm

When these macromolecules are oxidized they release energy, some of which is stored in the phosphate bonds of a small molecule, ATP (Adenosine triphosphate), which acts as a universal "energy currency"

ATP is recognized by the enzymes that run cell metabolism

ATP releases energy when one of the 2 terminal phosphates is split off (hydrolysis)

ATP → ADP + P + energy

The amount of energy released by ATP hydrolysis is of the right order of magnitude to do most cellular jobs: -7.3 kcal/mole (standard conditions); about -10 to -14 kcal/mole in the cell

Cells typically pump ions into and out of their interiors

This results in uneven distribution of ions on the 2 sides of the cell membrane

Example: in the body the external Na (in the blood) is about 150 mM, while the Na within cells is about 10 mM

Concentration differences like this are called ion gradients

Ion gradients represent stored electrical energy (batteries); the energy can be tapped by making it flow through biological machines

Example: mitochondria and chloroplasts make ATP by letting a H gradient flow through the enzyme ATP synthetase

Biochemical Reactions are Governed by Changes in Free Energy

Because there are so many molecules involved it is impossible to measure the kinetic energy of all the molecules or the electrical energy in each of the bonds

Energy changes in chemical reactions are analyzed from changes in chemical concentrations and the amount of heat produced

Analysis of these components (you will have to take a class in physical chemistry to see how this is done) shows that in spontaneous reactions disorder (entropy) always increases

Examples:

A hot object placed in the center of a room will cool and the rest of the room will be warmed (heat energy spreads out, less ordered)

A puff of smoke generated at one place in a room will again spread until it is evenly distributed throughout the room

Most biochemical reactions occur at constant temperature and pressure

Quantitative energy changes in biochemistry are measured by the free energy, G :

$G = H - TS$ = Gibbs free energy (this is used when the temperature is constant)

H = enthalpy = total energy of system; S = entropy (measure of disorder); T = temperature (in absolute degrees: degrees C + 273)

Biochemical reactions can occur if the energy stored in bonds decreases or if the disorder increases

H must decrease or S must increase or both such that G decreases

Change in G	What Happens
-	Reaction is spontaneous; will occur naturally
0	Reaction is at equilibrium; no changes
+	Reaction will not occur unless coupled with a second reaction with a large enough negative change in G

The Universe is Running Down but Life is Getting More Complex

The entropy of the universe is increasing as it runs down (Second Law of Thermodynamics)

All forms of energy are degraded to heat when they interact with matter

Heat flows from hot to cold temperatures until there is an even temperature everywhere

A system with only heat energy and an even temperature everywhere cannot do work because there is no way to make the heat flow

Example: The sun is running down, its energy is being dissipated, spreading throughout the universe

Life is getting more complex

Even simple cells are much more ordered than their surroundings

In evolution more and more complex organisms have evolved with time

These developments seem to contradict the Second Law of Thermodynamics

Chemical reactions can be coupled together so that one reaction drives the other

A reaction with a positive free energy can occur if it is coupled with another reaction with a large negative free energy

We exist and become more complex at the expense of other systems which are running down

In such coupled reactions the universe as a whole still runs down and the Second Law of Thermodynamics is not violated.

Living Organisms are not in Equilibrium With Their Surroundings

Equilibrium = death

Continuous input of energy is required to keep processes from going to equilibrium

To counteract the trend toward equilibrium cells burn fuel molecules (sugar, fat, amino acids) and use the released energy to make ATP

Cells exist in a stable relationsdhip with the environment called a steady-state

Example of steady-state 1: production of fresh water from sea water using energy from sun (non-living steady-state)

Example of steady-state 2: production of macromolecules from small molecules using ATP energy (living steady-state)

Example of steady-state 3: photosynthesis- production of sugars and other carbon compounds from CO_2 and water, using light energy (living steady-state)

In these steady states the system is pushed away from equilibrium by the continuous inflow of energy from outside

The universe as a whole is still running down (that's where the outside energy comes from); we become more complex at the expense of the universe

More than 99% of the energy for life comes from the sun

CHAPTER - 4

Enzymes and Kinetics

ENZYMES

Enzymes are the heart of Biochemistry

- Protein based catalysts.
- Enormously effective catalysts: typically enhance rates by 10^6 to 10^{12} fold.
- Operate under mild conditions: 0 - 100 °C (or perhaps even 300+ °C for some bacteria in deep ocean), 20 - 40 °C for most organisms; and low pressures (atmospheric).
- Very specific: Generally catalyze reaction for a very restricted group of molecules, sometimes for a single naturally occurring molecule of a single chirality.

Characteristics of Enzymes

1) Biological catalysts
2) Not consumed during a chemical reaction
3) Speed up reactions from 1000 - 10^{17},
4) Exhibit stereospecificity → act on a single stereoisomer of a substrate
5) Exhibit reaction specificity → no waste or side reactions

Classification of Enzymes

Main classification number	Major classes and sub-classes	Type of reaction catalysed
1.	**Oxidoreductases** Dehydogenases Oxidases Reductases Peroxidases Catalases Hydroxylases	Transfer of electrons usually in the form of hydrogen atoms or hydride ions A reduced + B oxidised → A oxidsed + B reduced CH3-CH2-OH + NAD+→CH3-CHO + NADH+ H^+ (reduced) (oxidised) (reduced)
2.	**Transferases** Kinases Acyltransferases Transaldolases Methyl transferases	Transfer of functional groups form one molecule to another. Example: Phosphorylation of glucose by hexokinase A B + C → A + B - C
3.	**Hydrolases** Esterases Glycosidases Peptidses Phosphatases Thiolases Phospholipases Amidases	Cleavage of bonds by hydrolysis A-B+ H_2O → A-H +B-OH
4.	**Lyases** Decarboxylases Aldolases Hydratases Dehydratases	Removal of groups by a mechanism other than hydrolysis leaving a double bond in one of the product. A - B → A = B + X - Y | | X Y
5.	**Isomerases** Racemases Epimerases Isomerases Mutases	Transfer of groups within a molecule to yield isomeric forms A - B → A - B | | | | X Y Y X
6.	**Ligases** Synthetases Carboxylases	Formation of new bonds(C-C, C-S, C-O C-N etc.) by condensation coupled with hydrolysis of high energy molecules like ATP). A B + C → A + B - C ATP ADP + Pi Glutamic acid + NH3 → Glutamine Enzyme: Glutamine synthetase

Comparision of apoenzymes, coenzymes and cofactors

A large number of enzymes require an additional non protein component to carry out its catalytic functions. Generally these non protein components are called as cofactors. The cofactors may be either one or more inorganic ions such as Fe^{2+}, Mg^{2+}, Mn^{2+} and Zn^{2+} or a complex organic molecules called coenzymes. Some enzymes require both coenzyme and one or more metal ions for their activity. A coenzyme or metal ion that is covalently bound to the enzyme protein is called prosthetic group. A complete, catalytically active enzyme together with its coenzyme and/or metal ions is called holoenzyme. The protein part of such an enzyme is called apoenzyme or apoprotein. Coenzymes function as transient carriers of specific functional groups

Enzymes requiring metal ions as cofactors

Enzymes	Cofactors
Cytochrome oxidase	Fe^{2+} and Cu^{2+}
Peroxidase, catalase	Fe^{2+} or Fe^{3+}
Carbonic anhydrase	Zn^{2+}
Alcohol dehydrogenase	Zn^{2+}
Hexokinase	Mg^{2+}
Pyruvate kinase	Mg^{2+}
Glucose 6-phosphatase	Mg^{2+}
Pyruvate kinase	K^{+}
Nitrogenase, nitrate reductase	Mo
Glutathione peroxidase	Se

Enzymes Kinetics and Inhibitors

Factors affecting enzymatic reaction

The factors that mainly influence any enzyme-catalysed reaction are

1. Substrate concentration
2. Enzyme concentration
3. Temperature
4. pH
5. Inhibitors

Other factors such as state of enzyme (oxidation), time and activators also affect enzyme-catalysed reaction to certain extent.

For enzyme-catalyzed reactions:

$$E + S \longrightarrow ES \longrightarrow E + P$$

The rate or velocity is dependent upon both [enzyme] and [substrate].

In reality, enzyme-catalysed reactions are not that simple:

$$E + S \underset{k_{-1}}{\overset{k_1}{\rightleftharpoons}} ES \xrightarrow{k_{cat}} E + P$$

k_1 and k_{-1} govern the rates of association and dissociation of ES

k_{cat} is the turnover number or catalytic constant

$$V_{ES} = k_1[E][S]$$
$$V_{E+S} = k_{-1}[ES]$$
$$V_{E+P} = k_{cat}[ES]$$

Usually an enzyme's velocity is measured under initial conditions of [S] and [P].

- At low [S], v_o increases as [S] increases.
- At high [S], enzymes become saturated with substrates, and the reaction is independent of [S] → display **saturation kinetics**.

$$V_{max} = k_{cat}[ES]$$

or because the [S] is irrelevant at high [S]

$$V_{max} = k_{cat}\,[E]$$

The graph is a graph of a hyperbola, and the equation for a hyperbola is

$$y = \frac{ax}{b + x}$$

where a is the asymptote
b is value at a/2

Substituting our equation parameters,

$$V_o = \frac{V_{max}[S]}{K_m + [S]}$$ **Michaelis-Menten equation**

Different enzymes reach V_{max} at different [S] because enzymes differ in their affinity for the substrate or Km.

1) The greater the tendency for an enzyme and substrate to form an ES, the higher the enzyme's affinity for the substrate → lower K_m.
2) At a given [S], the more enzyme will be in ES for an enzyme with a higher affinity

i.e. the greater the affinity, the lower the [S] needed to saturate the enzyme or to reach V_{max}.

Enzyme-substrate affinity and reaction kinetics are closely associated

[S] at which $v_o = 1/2V_{max} = K_m$

K_m is a measure of enzyme affinity

$$K_m = \frac{k_{-1}}{k_1}$$

reflection of association and dissociation of ES

- a small K_m (high affinity) favors $E + S \rightarrow ES$
- a large K_m (low affinity) favors $ES \rightarrow E + S$
- meaning that the lower the Km, the less substrate is needed to saturate the enzyme.

We would like numbers of V_{max} and K_m for a means of comparison among enzymes.

It is difficult to estimate V_{max} and K_m from a typical graph of [substrate] vs. velocity.

These two parameters are used to describe the efficiency of enzymes; must be an easier method for measuring these parameters.

Done by transformation of the date by taking the reciprocal of both sides of the equation —> double reciprocal plot or **Lineweaver-Burke plot**.

$$V_o = \frac{V_{max}[S]}{K_m + [S]}$$

$$\frac{1}{V_o} = \frac{K_m}{V_{max}}\frac{1}{[S]} + \frac{1}{V_{max}} \qquad \mathbf{y=mx+b}$$

Alterations in enzyme activity

Enzyme inhibition

Some compounds have the ability to combine with certain enzymes but do not serve as substrates and therefore block catalysis. These compounds are called inhibitors.

- Molecule that binds to enzyme and interferes with its activity to prevent either:
  1) Formation of ES complex E + I ® EI
  2) Breakdown of ES ® E + P ES + I ® ESI

- Used to regulate metabolism.
- Many drugs act by enzyme inhibition.
- These molecules can be

1) **Irreversible inhibitors** - bind to enzymes by covalent means and modify enzyme

   It bind to the enzyme and destroy the active site, or otherwise screw the protein. *Suicide inhibitors,* a special class of such inhibitors, are activated by the normal catalytic activity of the enzyme, but form an intermediate that binds to and destroys the active site. Irreversible inhibitors bind tightly (often covalently) to the enzyme and cannot be removed by dialysis. They include such things as nerve gases (Sarin, DIPF, Tabun) and insecticides (Malathion).

2) **Reversible inhibitors**- noncovalent binding to enzyme

There are three types of reversible inhibition:

1) Competitive inhibitors

   In *competitive inhibition,* both inhibitor and substrate can bind to enzyme and form two independent complexes. Only ES degrades to products: EI is considered a 'dead-end'. Because the inhibitor binds, to the active site, the substrate cannot (and *vice versa*), so there cannot be an ternary ESI complex.

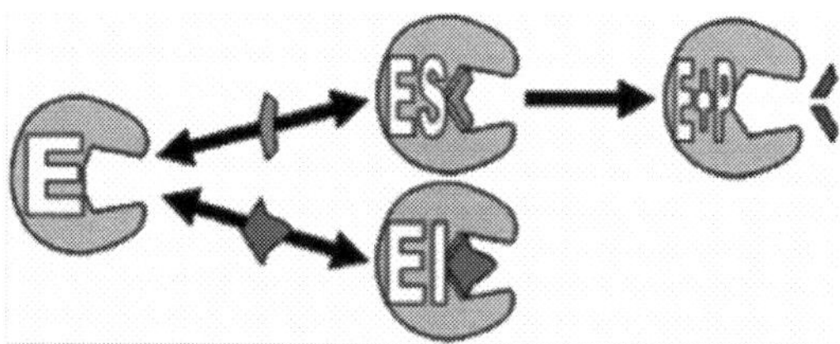

- Competes with substrate for active site of enzyme.
- Both substrate and competitive inhibitor bind to active site.
- These inhibitors are often substrate analogs (similar in structure substrate), but still no product is formed.
- Can be overcome by addition of more substrate (overwhelm inhibitor; a numbers game).

  e.g. malonate inhibition of succinate dehydrogenase

e.g. AZT inhibition of HIV reverse transcriptase actual substrate is dTTP (deoxythymidine triphosphate)

- Can be represented by the following equation:

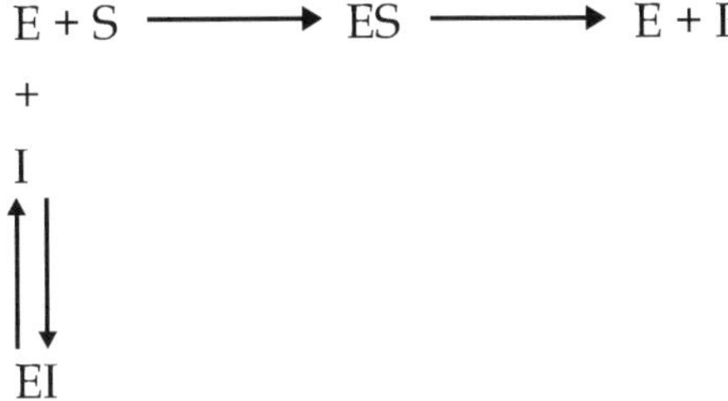

- affects K_m (increases K_m ® decreases affinity; need more substrate to reach half-saturation of enzyme)
- V_{max} unaffected

2) Uncompetitive inhibitor
   - Typically seen in multisubstrate reactions (here, there is a decrease in product formation because the second substrate cannot bind).
   - Inhibitor binds to ES, but not enzyme.

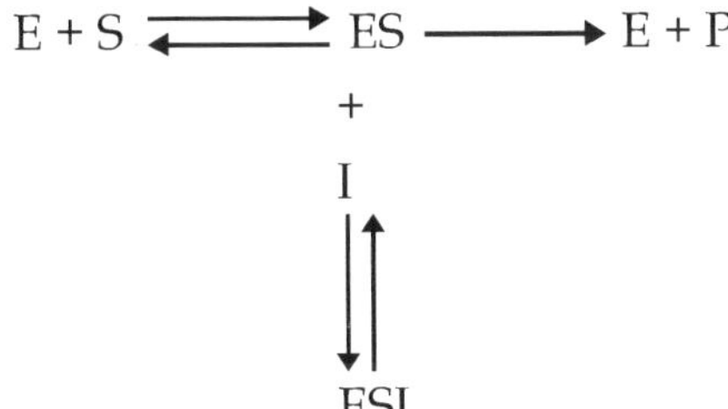

Graphical representation of uncompetitive inhibitors:

Lineweaver-Burke plot:

- both Km and V_{max} are lowered, usually the same amount
- ratio Km/V_{max} unchanged ® no change in slope

3) Pure noncompetitive inhibitor

Non-competitive inhibitors reversibly to somewhere other than the active site: they change the protein conformation allosterically, and reduce the rate at which the enzyme turns over product. They have no effect on K_m as the active-site of uninhibited enzyme molecules will only encounter substrate, and no unproductive binding will occur. They *do* however reduce the apparent V_m; consequently the apparent V_m will be reduced, since the protein is no longer as enzymatically competent. Such inhibitors are generally not substrate analogues.

Because the inhibitor can bind independently of the substrate, an ESI complex can also form. Both ESI and EI are dead-ends.

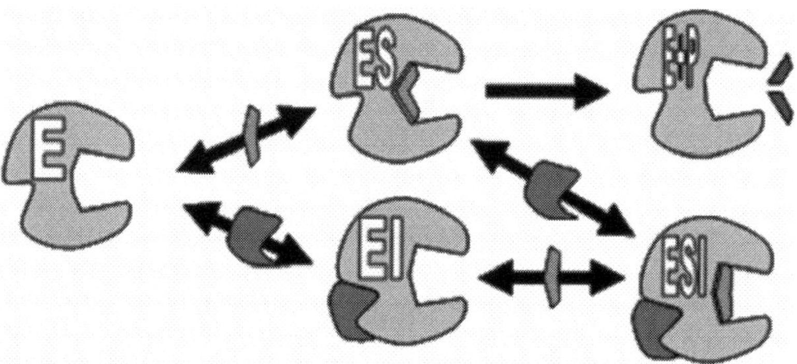

Non competitive inhibitors bind to a site other than the active site on the enzyme often to deform the enzyme, so that, it does not form the ES complex at its normal rate. Once formed, the ES complex does not decompose at the normal rate to yield products. These effects are not reversed by increasing the substrate concentration.

- Can bind to enzyme and ES complex equally.
- Does not bind to same site as substrate and is not a substrate analog.
- Cannot be overcome by increases in [substrate].

  e.g. lead, mercury, silver, heavy metals

Lineweaver-Burke plot

- No effect on K_m, because those enzyme molecules unaffected have normal affinity.
- V_{max} is lowered.

Regulation of Enzyme Activity

There are many ways to regulate enzyme activity at different levels:

1) **Regulation of rate of synthesis or degradation**

- Is fairly slow (several hours), so is really too slow to be effective in eucaryotic cells.
- Need something that can occur in seconds or less.
- Usually done through regulatory enzymes and occur in metabolic pathways early or at **first committed step**:

$$\overset{E_1}{A + B \rightarrow C} \rightarrow D \rightarrow E \rightarrow F \rightarrow P$$

Feedback inhibition

$$G \rightarrow H$$

- Result is to conserve material and energy by preventing accumulation of intermediates.

2) **Allosteric regulation**

- Done through **allosteric sites** or regulatory sites on enzymes - site other than active site where inhibitor or activator can bind.
- Properties of allosteric enzymes:
  1) sensitive to metabolic inhibitors and activators
  2) binding is noncovalent; not chemically altered by enzyme
  3) regulatory enzymes possess quaternary structure - individual polypeptide chains may or may not be identical
  4) enzyme has at least one substrate that gives sigmoidal curve due to positive cooperativity because of multiple substrate binding sites.

- Theories of allosteric regulation:

1) **Concerted theory or symmetry-driven theory**

Assumes 1 binding site/subunit for each ligand.

Enzyme can assume either R or T conformation.

Assumes that all subunits are in R or T state, and all switch at same time when the first substrate is bound.

2) **Sequential theory**

Ligand introduces a change in the tertiary structure of a subunit.

Only that subunit is converted to R conformation.

3) **Covalent modification**

- Usually requires one enzyme to activate enzyme and another to inactivate.
- Most common modification is phosphorylation of serine residues on interconvertible enzyme (the one that does the activating).

  e.g. pyruvate dehydrogenase

Enzyme Mechanisms

All enzymatic reactions go through a transition state (unstable intermediate form with a structure between that of reactant and product).

Reactants must collide precisely to form transition state.

Must have correct orientation.

Must collide with enough energy = activation energy = ΔG

Enzymes work by lowering ΔG

Substrates are correctly oriented. All increase probability of reaction.

Transition states are stabilized.

Chemical Catalysis

Active site of most enzymes is lined with hydrophobic amino acids.

There are a few polar a.a. which make up the catalytic center of the active site and can be ionized.

Histidine (basic a.a.) is common.

Aspartate and glutamate - negatively charged

Lysine and arginine - positively charged; electrostatic binding can occur

There are several types of chemical catalysis:

1) Acid-base catalysis

Enzymes that use this have a.a. side chains that can donate or accept electrons to substrate.

Can accelerate a chemical reaction by a factor of 10-100.

2) Covalent catalysis

Substrate forms a covalent bond with enzymes, then part of substrate is transferred to a second substrate in a 2 step process.

$$A\text{-}X + E \rightarrow X\text{-}E + A$$

$$X\text{-}E + B \rightarrow B\text{-}X + E$$

3) Proximity effect

Collection of substrate molecules in the active site increases the concentration over those molecules found freely in solution.

Result is that there is a more frequent formation of transition states.

4) Transition state stabilization

Increase binding of transition states to enzymes compared to substrate or product alone binding.

Binding forces are charge-charge interactions, hydrogen bonds, hydrophobic interactions, van der Waals forces.

Often seen in side chains of aspartate, glutamate, histidine, lysine, and arginine.

pH affects enzymatic rates:

Inflection points approximate pKa of ionizable residues important in active site.

e.g. papain

Three Examples of Enzymes Mechanisms

1) Triose phosphate isomerase

Catalyzes the interconversion of dihydroxyacetone phosphate (DHAP) and glyceraldehyde 3-phosphate.

Enzyme has two ionizable active site residues Glu (glutamate)-165 and His-95 → act as acid-base catalysts.

Mechanism:

1) Hydrogen bonds form between imidazole group of His-95 and carbonyl oxygen of DHAP (COO^- group is ionized; histidine active a.a.)

Carboxylate group of Glu-195 attacks proton of C-1 from substrate (DHAP) to form enediolate intermediate.

2) C-2 oxygen (electron-rich) attacks proton of His-95 → converts - O^- to -OH to form an enediol.

3) imidazole from of His-95 attacks -OH of C-1 to form another enediolate intermediate

4) Glu-195 donates proton to C-2 → glyceraldehyde 3-phosphate formed

2) Lysozyme

Good example of transition state stabilization - enzyme binds to and stabilizes transition state.

Action: hydrolyzes bacterial cell walls. Found in tears, saliva.

Substrate is alternating residues of N-acetylglucosamine (GlcNAc) and N-acetylmuramic acid (MurNAc).

Enzyme hydrolyzes glycosidic bond.

Binding site can hold 6 sugars or residues labeled A, B, C, D, E, F - one of the residues (D) must be bent into half-chair conformation.

Glu-35 and Asp-52 are the two a.m. residues most involved.

Mechansism:

1) MurNAc residues bind at B, D, F → residue D turned into half-chair conformation.

2) Glu-35 acts as acid catalyst → donates H^+ to O of glycosidic bond between D and E → glycosidic bond broken

3) Residues bound in E and F diffuse out of binding site.

4) Replaced by a molecule of water.

5) C-1 now is a carbonium ion intermediate C^+.

6) Asp-52 stabilizes carbonium intermediate via charge-charge interactions.

7) Glu-35 O^- attacks H^+ from water molecule —> OH^- is added to C^+ ion

8) R-MurNAc diffuses away (enzyme has been reprotonated; no charges to hold it in place).

3) Chymotrypsin, a member of the serine proteases

Examples: trypsin, chymotrypsin, and elastase that catalyze much of the digestion in the small intestine.

Synthesized in pancreas; stored as inactive precursors called zymogens - prevents damage to cell.

What is actually released from the cell is trypsinogen, chymotrypsinogen, proelastase, which are then activated by selective proteolysis.

Enzyme called enteropeptidase activates trypsinogen → trypsin by cleaving of N-terminal hexapeptide.

Trypsin then activates the other two.

All three enzymes have similar primary, second, tertiary structure.

All cleave peptide bonds on COOH side of hydrophobic or aromatic side chains.

Substrate specificity is due to amino acid residues in the hydrophobic binding pocket.

chymotrypsin - serine (uncharged) → accepts large, bulky, hydrophobic side groups.

trypsin - aspartate (negatively charged) accepts Lys, Arg, Gly, Ala.

elastase - shallow - binds a.a. with small side chains (Gly, Ala).

Catalysis involves use of catalytic triad:

Ser-195 His-57 Asp-102

All three amino acids are hydrogen bonded.

Ser-195 residue is highly reactive; very unusual.

Mechanism of chymotrypsin:

1) Substrate enters enzyme and is aligned with R1 group in binding pocket → places carbonyl carbon of peptide bond next to oxygen of Ser- 195.
2) His-57 attacks H of Ser-195.
3) Now, the nucleophilic oxygen of Ser-195 attacks carbonyl carbon of peptide bond to form tetrahedral intermediate (transition state?).
4) C=O bonds changes to a single bond (oxygen is negatively charged = oxyanion) and forms H-bond with -NH groups of Gly-193 and Ser-195.
5) His-57 and Asp-102 share H^+ in **low-barrier hydrogen bond** (increases rate of catalysis by decreasing activation energy; very strong hydrogen bond).
6) His-57 imidazolium ring acts as an acid catalyst by donating H to peptide bond → molecule cleaved → amine product released.

7) Carbonyl group of peptide forms covalent bond with enzyme → acyl-enzyme intermediate formed.

8) After first product leaves, a molecule of water enters → donates H^+ to His-57 → -OH group left attacks carbonyl group → formation of second tetrahedral intermediate and stabilized by oxyanion hole plus a low barrier H-bond.

9) His-57 donates a proton → second intermediate collapses

10) Second product is formed, released from active site → chymotrypsin regenerated.

Shows: covalent catalysis (Ser), acid-base catalysis (His)

CHAPTER - 5

Coenzymes and Vitamins

There are other groups that contribute to the reactivity of enzymes beside amino acid residues.

These groups are called **cofactors** - chemicals required by apoenzymes (inactive) to become holoenzymes (active).

There are two types of cofactors:

1) **Essential ions** - metal ions -inorganic
2) **Coenzymes** - organic molecules that act as group-transfer reagents (accept or donate groups)- can also be H^+ and/or e^-

Both provide reactive groups not found on amino acids side chains.

Coenzymes can be either **cosubstrates** (loosely bound to enzyme; is altered, then regenerated) or **prosthetic groups** (tightly bound to enzyme).

Coenzymes can be classified by their source:

1) **Metabolite coenzymes**
   - synthesized by common metabolites
   - include nucleoside triphosphates
   - most abundant is ATP, but also include uridine diphosphate glucose (UDP-glucose) and S-adenosylmethionine
   - ATP can donate all of its three phosphoryl groups in group-transfer reactions
   - S-adenosylmethionine can donate its methyl group in biosynthetic reactions.

- UDP-glucose is a source of glucose for synthesis of glycogen in animals and starch in plants.

2) **Vitamin-derived coenzymes**

- Vitamins are required for coenzyme synthesis and must be supplied in the diet
- Lack of particular vitamins causes disease
- There are two catagories of vitamins:
    1) water-soluble - B vitamins and vit. C

       required daily in diet

       excess excreted in urine
    2) lipid-soluble - vitamins A, D, E, K

       Intake must be limited

       Stored in fat

VITAMINS AND THEIR COENZYMES

Vitamins are low molecular weight organic compounds required in small amounts in the diet. Most of the vitamins are not synthesized in the human body but are synthesized by the plants. Hence these essential nutrients are mainly obtained through the food. Though most of them are present in the diet as such, some are present as precursors known as provitamins.

Vitamins are divided into two major categories. They are fat-soluble (A, D, E and K) and water-soluble vitamins (B-complex and vitamin C).

B complex vitamins include thiamine (B_1), riboflavin (B_2), pantothenic acid (B_3), niacin (B_5), pyridoxine (B_6), biotin (B_7), folic acid (B_9), and cobalamin (B_{12}).

Inositol, cholic and para-aminobenzoic acid are vitamin-like substances sometimes classified as part of the B complex, but no convincing evidence has been shown so far to be included as vitamins.

All the fat-soluble vitamins and some B vitamins exist in multiple forms. The active forms of vitamin A are retinol, retinal and retinoic acid and vitamin D is available as ergocalciferol (D_2) and cholecalciferol (D_3).

The vitamin E family includes four tocopherols and four tocotrienols but a-tocopherol being the most abundant and active form

The multiple forms of vitamins are interconvertible and some are interchangeable.

Niacin (nicotinic acid) (B_5) $\rightarrow$ nicotinamide $\rightarrow$

Get niacin in enriched cereals, meat, legumes.

NAD^+ and $NADP^+$ are the coenzymes (cosubstrates).

NAD^+ consists of 2 5′ribonucleotides (AMP and nicotinamide monomucleotide) joined by a phosphoanhydride linkage.

For $NADP^+$, have a phosphoryl group on 2′-oxygen.

Both coenzymes act as cosubstrates for dehydrogenases → catalyze the oxidation of substrates by transfer of $2e^-$ and $1H^+$ → NADH and NADPH.

Vitamin B_2 (Riboflavin)

Coenzymes are flavin mononucleotide (FMN) and flavin adenine dinucleotide (FAD).

Riboflavin found in milk, whole grains, liver.

The coenzymes serve as prosthetic groups involved in $1e^-$ or $2e^-$ transfers.

$FAD + 2e^- + 2H^+ \rightarrow FADH_2$

$FMN + 2e^- + 2H^+ \rightarrow FMNH_2$

Enzymes that require FAD or FMN are called **flavoenzymes** or **flavoproteins**.

Can actually donate 1 or 2 e^- at a time → form partially oxidized compound when only $1e^-$ is donated → relatively stable.

Vitamin B_1 (Thiamine)

Structure: pyrimidine ring and positively charged thiazolium ring.

Found in husks of rice and other cereals, liver, meat, particularly pork.

Deficiency in thiamine causes beriberi - extensive nervous system and circulatory system damage, muscle wasting, edema.

Coenzyme form is thiamine pyrophosphate (TPP) - synthesized by transfer of pyrophosphoryl group from ATP via thiamine pyrophosphate synthetase.

Used primarily in decarboxylases as a coenzyme.

Vitamin B6 Family

pyridoxine, pyridoxal, pyridoxamine are the vitamins.

Act as prosthetic groups.

Formed by the following reaction:

pyridoxine + ATP → pyridoxine 5′phosphate → pyrodoxal 5′ phosphate (PLP).

Lack of B_6 results in defects in protein metabolism.

PLP found in enzymes that catalyze reactions involving amino acids, e.g. isomerizations, decarboxylations, R-group removal or replacements.

Most frequent reaction is a transamination, where the a-amino group of a.a. is transferred to carbonyl group of a-keto acid → new a.a. made or is excreted.

PLP binds covalently with Lys residue in active site → keeps PLP from running away.

Biotin

Synthesized by intestinal bacteria.

Prosthetic coenzyme is called biocytin - covalently linked to Lys residue in active site.

Involved in carboxyl group transfer reactions and ATP-dependent carboxylations.

E.g. pyruvate carboxylase

pyruvate + HCO_3^- → oxaloacetate

Binds to HCO_3^- and acts as a CO_2 carrier.

Folic Acid or Folate

Found in green leafy vegetables, liver, yeast.

Coenzyme form is tetrahydrofolate.

Used by enzymes that transfer 1-C units as methyl groups (CH_3^-).

Another folate coenzyme is tetrahydrobiopterin - used in hydroxylases.

Pantothenic Acid

Used in coenzyme A formation.

Reactive center is -SH group

Key in all acyl-group transfers

Coenzyme form is phosphopantethine - added to serine residue of protein → acyl carrier protein (ACP) → important in fatty acid synthesis.

Vitamin B_{12} or Cobalamin

Found in organ meat (kidney and liver).

It is a prosthetic coenzyme.

Ring structure similar to heme, with cobalt atom in center.

Involved in molecular rearrangements.

Deficiency in B12 results in pernicious anemia (decreased production of blood cells from bone marrow).

Vitamin C or Ascorbic Acid

Found in fresh fruit and vegetables.

Participates in hydroxylation reactions, e.g. collagen synthesis.

Deficiency causes scurvy.

Lipid Vitamins

Vitamin A or retinol

Is a 20 carbon lipid molecule.

Found in carrots, yellow vegetables, liver, egg yolk, milk products.

-carotene → vitamin A

Exists in three forms:

1) retinol and 2) retinoic acid - binds to intracellular protein receptors ® regulates gene expression

2) retinal - prosthetic group of rhodopsin

Vitamin D

Exists as several lipids;

1) D3 - made in skin exposed to sunlight.

2) D2 - additive in fortified milk

Deficiency causes rickets in children or osteomalacia in adults → insufficient Ca phosphate deposition in bone.

Vitamin E or a - tocopherol

Is an antioxidant that scavenges free radicals.

Vitamin K or phylloquinone

Found in plants.

Required for synthesis of proteins involved in blood coagulation.

Ubiquinone or coenzyme Q

Ring with hydrophobic tail → inserted into membranes.

Transports e^- between enzyme complexes in inner mitochondrial membrane.

Related molecule is plastiquinone - found in thylacoid membrane of chloroplasts.

Cytochromes

Hemo-containing protein coenzyme $Fe^{3+} \leftrightarrow Fe^{2+}$.

Classified as a, b, c based on absorption spectra.

Transfers e^-.

Coenzymes derived from vitamins

Coenzyme	Vitamin Precursor
Coenzyme A	Pantothenic acid
Thiamine pyrophosphate	Thiamine (B1)
Tetrahydrofolate	Folic acid
NAD, NADP	Niacin
FAD, FMN	Riboflavin (B2)

Vitamins, Their Dietary Sources and Deficiency Symptoms In Humans

Water-soluble vitamins

Vitamin	Some common dietary sources	Deficiency symptoms in humans
Thiamine (Vitamin B_1)	Liver, meat, milk, vegetables, whole grains, nuts	Dry and wet beri-beri. Weight loss-, muscle wasting, sensory changes, mental confusion, enlargement of heart, constipation
Riboflavin (Vitamin B_2)	Liver, wheat germ, eggs, milk, green leafy vegetables, meat	Magenta-coloured tongue, fissuring at the corners of mouth and lips, dermatitis
Pantothenic acid (Vitamin B_3)	Eggs, peanuts, liver, meat, milk, cereals, vegetables	Vomiting, abdominal distress, cramps, fatigue, insomnia
Niacin or nicotinic acid (Vitamin B_5)	Meat, liver, cereals, legumes	Pellagra. Dermatitis when exposed to sunlight, weakness, insomnia, impaired digestion, diarrhea, dementia, irritability, memory loss, headaches
Pyridoxine or pyridoxol (Vitamin B_6)	Egg yolk, fish, meat, lentils, nuts, fruits, vegetables	Convulsions, dermatitis, weight loss, irritability, weakness in infants
Biotin (Vitamin B_7)	Liver, yeast, meat, peanuts, eggs, chocolate, dairy products, grains fruits, vegetables	Dermatitis, skin dryness, depression, muscle pain, nausea, anorexia (appetite loss)

Contd...

Folic acid (Vitamin B_9)	Yeast, liver, green vegetables, some fruits	Anemia leading to weakness, tiredness, sore tongue, diarrhea, irritability, headache, heart palpitations
Cobalamin (Vitamin B_{12})	Meat, shellfish, fish, milk, eggs	Neurological disorders, anemia leading to tiredness, sore tongue, constipation, headache, heart palpitations.
Ascorbic acid (Vitamin C)	Vegetables and citrus fruits	Sore gums, loose teeth, joint pain, edema, anaemia, fatigue, depression, impaired iron absorption, impaired wound healing.

Fat soluble vitamins

Vitamin	Functions	Some common dietary sources	Deficiency symptoms
Vitamin A	Visual cycle and maintaining epithelial cells	Fruits, vegetables, fish-liver oils	Night blindness and eventually total blindness, anorexia (appetite loss), dermatitis, recurrent infections; in children, cessation of skeletal growth and lesions in the central nervous system
Vitamin D	Calcium metabolism	Fish-liver oil	Bone pain and skeletal deformities such as bowlegs (Rickets) and knock-knee in children. Osteomalacia in adults.
Vitamin E	Antioxidant	Plant oils, green leafy vegetables, milk, eggs, meat	Symptoms in humans, if any, are controversial; possibly anaemia
Vitamin K	Blood clotting	Leafy vegetables, soybeans, vegetable oils	Impaired blood clotting

CHAPTER - 6

Cell Structure and Their Organelles

ALL LIVING ORGANISMS ARE MADE UP OF UNITS CALLED CELLS

All living creatures are made from 1 or more cells

All cells are produced from previously existing cells (no spontaneous generation)

All cells appear to be descended from the first cell which existed about 4 billion years ago

For a species to exist its reproductive cells must be potentially immortal (no aging)

Our bodies start from a single cell and contain about 100,000,000,000,000 (10^14) cells at maturity

There Are 2 Basic Types of Cells: Prokaryotic and Eukaryotic

Prokaryotic cells are more primitive, small and without organelles: bacteria, cyanobacteria (blue-green algae)

Eukaryotic cells are more advanced, larger, contain organelles: all higher species (animals, plants, fungi, protozoa)

Our cells are of the eukaryotic type

The Surface/Volume Ratio of Cells Decreases as They Get Larger

For a spherical cell:

Surface Area = $4\pi r^2$
Volume = $(4/3)\pi r^3$
Surface Volume Ratio = $3/r$

Similar equations hold for other shapes

The surface volume ratio controls the metabolic rate and other functions associated with metabolism

Cell needs for food, energy, etc., are proportional to the volume

Rate at which needs can be supplied is proportional to the surface area

If delivery of vital substances is slow cell must slow down its metabolism

Small Prokaryotic Cells are Simple but Fast

Small cells such as bacteria divide fast (~ 20 min)- due to the high surface/volume ratio)

Size: mycoplasmas : 0.1-1 micron dia.; other bacteria 1-10 micron dia.

Have no nucleus: DNA less protected, mutates faster

One compartment- like a chem lab with a single beaker

Most have a rigid cell wall

May have a capsule outside the cell wall, pili (for attachment), flagella (for motion)

Large Eukaryotic Cells are Slow but Versatile

Cell division very slow (~ 20 hours)- due to the low surface/volume ratio)

Size: typically 10-100 micron dia.; volumes typically 1000 to 1 million times larger than prokaryotes

Have nucleus: DNA better protected, slow mutation rate

Have many organelles, most surrounded by bilayer lipid membranes

Organelles allow many activities to take place within the same cell- like a chem lab with many beakers; other reactions take place on membrane surfaces and eukaryotic cells have much more internal membrane surface that prokaryotic cells

Relative volumes of compartments in a liver cell:

Organelle	% Total Volume	Approx. Number per Cell
Cytosol	54	1
Mitochondria	22	1700
Rough Endoplasmic Reticulum	9	1
Smooth Endoplasmic Reticulum + Golgi	6	
Nucleus	6	1
Peroxisomes	1	400
Lysosomes	1	300
Endosomes	1	200

Data from: Alberts, *et al.* Molecular Biology of the Cell, 1994.

Some of the Organelles Found in Eukaryotic Cells Come From Endosymbiosis

Cells often ingest other cells and digest them for food

Sometimes the ingested cell is not digested, but the 2 cells learn to live together for mutual benefit (endosymbiosis)

Mitochondria and plant chloroplasts are believed to have originated in this way- have their own DNA and double membranes (2 bilayers)

The Cell Organelles are Found within the Cytosol

Cytosol is the liquid matrix of the cell- a watery solution containing a lot of protein

Contains salts, dissolved molecules, enzymes, etc.

Glycolysis takes place in cytosol

Gives 2 ATPs per glucose (does not require O2)

Cytoplasm = cytosol + organelles - nucleus

The Cell Membrane Separates the Cytoplasm From the External World

A lipid bilayer (2 molecular layers) with proteins inserted through it

Barrier to movement of things in and out of the cell (permeability)- hydrophobic molecules pass through it more readily than hydrophilic ones

Very specialized transport mechanisms

Supported on inside by protein filaments (cytoskeleton)

The Cytoskeleton Determines the Shape of the Cell

Three types of tubules are imbedded in the cytosol

Form a meshwork of fibers that:

- Give the cell shape
- Are used to transport structures within the cell (i.e., chromosomes in mitosis)
- Are involved in movement of the whole cell

Three basic types of fibers:

- Microtubules (made of tubulin, 25 nm dia)
- Intermediate filaments (made of several proteins, 8-12 nm dia)
- Microfilaments (made of actin, 7 nm dia)

Plant Cell Walls Give Rigidity to the Plant

Plant cells and bacteria have a cellulose wall outside of the cell membrane

- Has little effect on cell permeability but gives rigidity to cells
- Like an external skeleton for plant cells
- Gives protection against osmotic lysis

The Nucleus Contains the Molecule of Heredity: DNA

Contains the DNA (genetic information)- DNA does not leave nucleus

DNA associated with protein:

- Turns genes on and off
- Packages DNA for cell division

Repair mechanisms for DNA

Site of transcription of DNA (makes RNA copy of gene)

May contain 1 or more nucleoli (for making ribosomes)

Surrounded by membranes with special pores that let RNA out

The Centrioles Organize the Mitotic Spindle for Cell Division

Centrioles are a pair of small structures seen near the nucleus

Divide before cell division

Organize the mitotic spindle (made of tubulin) used in cell division

Contain a set of 9 triplet tubules

The Mitochondria are the Powerhouses of the Cell

Sites of cell respiration

Requires oxygen

Produce 36 ATPs/glucose molecule- major source of cell energy

Covered by 2 bilayer membranes

Large organelle cylindrical organelle, about the size of a bacterium (about 1 micron length)

Cells have about 1000, active cells like muscle have more

Probably originated from bacteria by endosymbiosis

Have small amounts of DNA (left over from when they were independent microorganisms?)

All your mitochondria come from your mother (very few in sperm)

Chloroplasts Use Energy from Sunlight to Make Sugar from CO_2 and Water

Plant cells have chloroplasts

Convert CO_2 and water to sugar, using light energy from the sun (photosynthesis)

Produce O_2 as a waste product- source of all O_2 in air

Very large, about 5 micron length

Like mitochondria have double bilayer membrane

Have own DNA, but most chloroplast proteins are coded by nuclear genes, not chloroplast genes

Proteins are Made on the Ribosomes

Ribosomes are made in nucleolus, then leave nucleus and enter cytoplasm

Made of RNA and protein

Some ribosomes are free in the cytosol, others attach to the ER

Each has 2 subunits

Decode the genetic code and make protein (translation)

Some Ribosomes Attach to the Endoplasmic Reticulum

Endoplasmic reticulum (ER) made up of lipid membranes

Some ER has attached ribosomes = rough ER

ER without ribosomes = smooth ER

Proteins that are secreted by the cell or which go to other organelles are made on the rough ER; cytosol proteins are made chiefly on free ribosomes

Rough ER highly developed in secretory cells such as those of liver, pancreas

The Smooth ER Makes Lipids and Detoxifies Drugs

Smooth ER, with no ribosomes, is involved in lipid metabolism- makes cell membranes, steroid hormones, etc.

Liver smooth ER has enzymes that detoxify drugs (cytochrome P450 system)

In muscle special smooth ER accumulates Ca ions (trigger for muscle contraction)

Proteins are Finished off and Routed in the Golgi Apparatus

Golgi apparatus is a set of stacked membranes found near the nucleus

Golgi finishes proteins: makes additions of small polysaccharides, cuts some of them, etc.

Routes proteins to right destination: some go to mitochondria, others to lysosomes, still others to cell membrane, etc.

Lysosomes Digest Materials within the Cell

Small vesicles surrounded by membranes

Lysosomes contain digestive enzymes that break down proteins, lipids, etc

Break down defective cell parts so they can be recycled

Also digest food brought into cell by phagocytosis

Involved in apoptosis (programmed cell death)

Require an acid pH inside (~5)

Endosomes Contain Large Proteins Brought into Cell by Endocytosis

Endosomes are food vacuoles

Brought into cell by endocytosis

Fuse with lysosomes for digestion

Peroxisomes Deal with Reactive Oxygen Molecules Such as Peroxides

Contain the enzyme, catalase, which converts hydrogen peroxide to O2 and water

Another enzyme, urate oxidase, sometimes forms crystals within the peroxisome

Important in fat metabolism

Cilia and Flagella Allow Cells to Move

Eukaryotic cilia & flagella are whiplike projections from the cell

- Have same internal structure: 9 pairs of tubules arranged in circle, surrounding a central pair of tubules
- Beat repetitively (a bending motion) and cause cell to move (or move fluids along a surface of cells)
- Bending caused by a contractile protein, dynein
- Enclosed within the cell membrane
- Made of at least 200 different proteins
- Some biologists call them undulipodia

Bacterial flagellum is very different

- Does not have 9 + 2 structure
- Made of single protein, flagellin
- Rotates through a shaft in cell membrane/wall

Vacuoles are Storage Compartments of Cells

Cells contain small vacuoles ("containers"), used for storing material such as glycogen and lipids

Plant cells have an enormous vacuole, filling most of the cell

- Surrounded by membrane called the tonoplast
- Stores water soluble materials such as anthocyanin pigments, opium, garlic flavor, proteins, tannins, etc.

Some protozoa (i.e., ameoba) have a contractile vacuole

- When the cell takes in too much water (osmosis) the vacuole lining contracts, squirting out excess water

Defective Cell Organelles are Responsible for Some Diseases

Examples:

a) Lysosomal storage diseases such as Tay-Sachs: lipids accumulate in lysosomes because they cannot be broken down

b) Cilia paralyzed by tobacco smoke and other pollutants cannot move mucus. Mucus accumulates in the lungs, impairing respiration

c) Lactic acidosis- can be produced by abnormal mitochondria with defective aerobic metabolism. In this situation lactic acid can accumulate in the blood.

Eukaryotic Cell Organelles

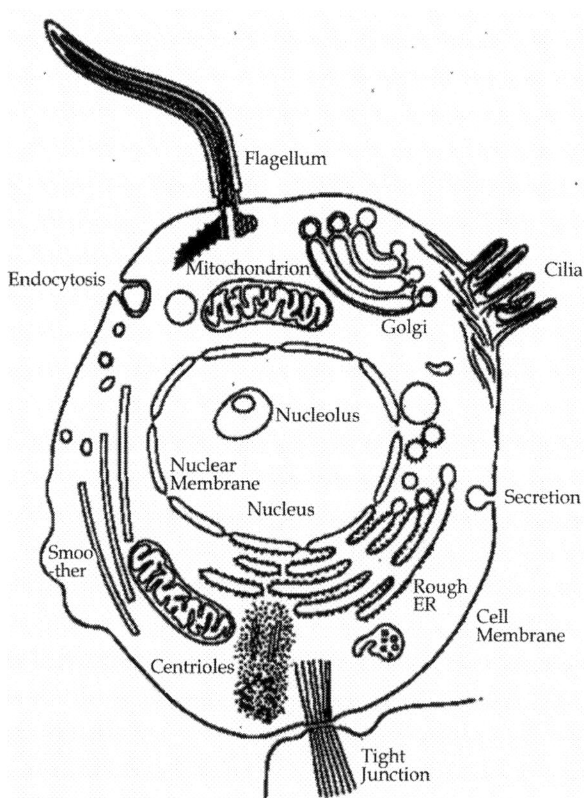

Different major metabolic pathways occur in a "typical" **liver cell.**

- **Endoplasmic Reticulum**: Lipid Synthesis; Steroid synthesis; Phase one detoxification reactions; Biosynthesis and modification of membrane and export proteins.

- **Plasma Membrane**: Active and passive transport systems; receptors and signal processing systems (synthesis of various second messengers etc.).
- **Nucleus**: DNA replication, synthesis and processing of messenger RNA's.
  - **Nucleolus**: localized region of the nucleus in which ribosomal RNA's are synthesized and processed.
- **Cytosol**: Glycolysis and most of gluconeogenesis; Pentose Phosphate shunt; Fatty acid biosynthesis.
- **Mitochondria**: Kreb's Citric Acid Cycle; Electron transport system and Oxidative Phosphorylation; Fatty acid oxidation; Amino acid catabolism; Interconversion of carbon skeletons.
- **Golgi Complex**: Further modification of membrane and export proteins.
- **Lysosomes**: Hydrolytic (digestive) enzyme localization.
- **Peroxisomes**: Amino acid oxidases, catalase-oxidative degradation reactions.
- **Glycogen Granules**: enzymes of glycogen synthesis and breakdown. including branching and debranching.

There are several types of plastids found in plant cell:

- **Proplastids** - small, precursors to the other plastid types, found in young cells, actively growing tissues;
- **Chloroplasts** - sites of photosynthesis (energy capture). They contain photosynthetic pigments including chlorophyll, carotenes and xanthophylls. The chloroplast is packed with membranes, called thylakoids. The thylakoids may be stacked into pancake- like piles called grana (granum, singular). The "liquidy" material in the chloroplast is the stroma. A chloroplast is from 5-20 ¼m in diameter and there are usually 50-200 per cell. The chloroplast genome has about 145 Kbase pairs, it is smaller than that of the mitochondria (200 kbases). About 1/3 of the total cell DNA is extranuclear (in the chloroplasts and mitochondria);
- **Chromoplasts** - non-photosynthetic, colored plastids; give some fruits (tomatoes, carrots) and flowers their color;
- **Amyloplasts** - colorless, starch-storing plastids;

- **Leucoplast** - another term for amyloplast;
- **Etioplast** - plastid whose development into a chloroplast has been arrested (stopped). These contain a dark crystalline body, prolamellar body, which is essentially a cluster of thylakoids in a somewhat tubular form.

Plastids can dedifferentiate and convert from one form into another. For example, think about the ripening processing in tomato. Initially, green tomatoes have oodles of chloroplasts which then begin to accumulate lycopene (red) and become chromoplasts. Usually you find only chromoplasts or chloroplasts in a cell, but not both.

Organelles of eukaryote and its key function in the cell

Organelle	Structure/Function
Cell Membrane	The cell membrane keeps the cell together by containing the organelles within it. Cell membranes are selectively-permeable, allowing materials to move both into and outside of the cell.Active and passive transport systems; receptors and signal processing systems (synthesis of various second messengers etc.).
Centrosomes	The centrosomes contain the centrioles, which are responsible for cell-division.
Cytoplasm	Cytoplasm is a jelly-like substance that is sometimes described as "the cell-matrix". It holds the organelles in place within the cell. Site for Glycolysis and most of gluconeogenesis; Pentose Phosphate shunt; Fatty acid biosynthesis.
Golgi Apparatus	The gogli apparatus of a cell is usually connected to an endoplasmic reticulum (ER) because it stores and then transports the proteins produced in the ER.Further modification of membrane and export proteins.
Glycogen Granules	Enzymes of glycogen synthesis and breakdown including branching and debranching.
Lysosomes	Lysosomes are tiny sacs filled with enzymes that enable the cell to utilize its nutrients. Lysosomes also destroy the cell after it has died, though there are some circumstances (diseases/conditions) in which lysosomes begin to 'break-down' living cells.Hydrolytic (digestive) enzyme localization.
Microvilli	"Microvilli" is the pural form; "Microvillus" is the singular form.Microvilli are finger-like projections on the outer-surface of the cell. Not all cells have microvilli. Their function is to increase the surface area of the cell, which is the area through which diffusion of materials both into, and out of, the cell is possible.
Mitochondria	"Mitochondria" is a plural term; which is appropriate as these are not found alone. The quantity of mitochondria within cells varies with the type of cell.These are the energy producers within the cell.

Contd...

	They generate energy in the form of Adenosine Tri-Phosphate (ATP). Generally, the more energy a cell needs, the more mitochondria it contains.Site for Kreb's Citric Acid Cycle; Electron transport system and Oxidative Phosphorylation; Fatty acid oxidation; Amino acid catabolism; Interconversion of carbon skeletons.
Nuclear Membrane	The nuclear membrane separates the nucleus and the nucleolus from the rest of the contents of the cell.
Nuclear Pore	Nuclear pores permit substances (such as nutrients, waste, and cellular information) to pass both into, and out of, the nucleus.
Nucleolus	The nucleolus is responsible for the cell organelles (e.g. lysosomes, ribosomes, etc.).Localized region of the nucleus in which ribosomal RNA's are synthesized and processed.
Nucleus	The nucleus is the "Control Center" of the cell, which contains DNA (genetic information) in the form of genes, and also information for the formation of proteins.Information is carried on chromosomes, which are a form of DNA.Site for DNA replication, synthesis and processing of messenger RNA's.
Peroxisomes	Amino acid oxidases, catalase-oxidative degradation reactions.
Ribosomes	Ribosomes interpret cellular information from the nucleus and so synthesize appropriate proteins, as required.
Rough Endoplasmic Reticulum (RER)	"Rough" indicates that there are ribosomes attached to the surfaces of the endoplasmic reticulum. The endoplasmic reticulum is the site for membrane and secretary protein biosynthesis.
Smooth Endoplasmic Reticulum (SER)	"Smooth" indicates that there are no ribosomes attached to the surfaces of the endoplasmic reticulum. Site of phospholipid biosynthesis and detoxification reactions takes place.
Chloroplasts	Light capturing processes and electron transport & oxidative phosphorylation for photosynthesis; Calvin cycle (dark reactions of photosynthesis).
Glyoxisomes	Location of glyoxalate cycle.
Cell wall	Made up of cellulose glued together with lignin (a plastic like polymer) - maintains cell integrity against high osmotic pressure, gives cell rigidity.
Vacuole	Storage of dilute aqueous solutions, provides fluid for osmotic pressure.

Differences between Prokaryotic and Eukaryotic cell.

	Prokaryotic cell		Eukaryotic cell
1.	Nucleus is incipient type. It lacks nuclear membrane, nucleolus and chromatin reticulum (chromosome).	1.	It has a true nucleus in which nuclear membrane, nucleolus and chromatin reticulum are present.
2.	DNA is circular and non-histone type, i.e. lacks protein association and lies in a tangled mass called nucleoid.	2.	DNA is associated with protein to form chromosomes or chromatids.
3.	All membrane-bounded structures such as chloroplast, mitochondria, ER, Golgi complex, vacuoles etc. are absent.	3.	membrane-bounded structures are present.
4.	Cell wall contains aminosugars and muramic acids.	4.	These substances are absent in the cell wall.
5.	Cytplasmic streaming (cyclosis) is not observed.	5.	It can be observed.
6.	In autotrophic forms, photosynthetic pigments are found associated with lamellae but not enclosed by membranes.	6.	Lamellae are enclosed in two unit membrane envelope and called chloroplast.
7.	Flagella do not show 9+2 fibril arrangement. These are hollow and tubular.	7.	Flagella comprise of 2 central and 9 peripheral fibrils.
8.	Mesosomes are found.	8.	Mitochondria are found instead of mesosomes.
9.	Cell wall and cell membrane have intricate contact and these do not fully separate during plasmolysis.	9.	Cell membrane lies separate and show complete shrinkage during plasmolysis.
10.	Endocytosis and exocytosis are not found	10.	Both phenomena are found.
11.	Only 70s type of ribosomes are found.	11.	Both 70s and 80s ribosomes are found.
12.	Microtubules and nucleosomes are not found.	12.	Microtubules and nucleosomes are found.
13.	RNA plays major role in condensation of DNA forming genophore.	13.	Proteins (histones and hestones) play major role in condensation of DNA forming chromatid or chromosomes.
14.	A single replication site is formed during replication of DNA.	14.	Several replication sites are formed during DNA replication.
15.	During transcription, a single type of RNA polymerase is found.	15.	During transcription, a three type of RNA polymerase is found.
16.	nif-genes are found in some prokaryotes.	16.	These are not found in eukaryotes.

Chapter - 7

Cell Wall - Structure and Functions

I. WALL COMPONENTS – CHEMISTRY

The main ingredient in cell walls are polysaccharides (or complex carbohydrates or complex sugars) which are built from monosaccharides (or simple sugars). Eleven sugars are common in these polysaccharides including like glucose and galactose. Carbohydrates are good building blocks because they can produce a nearly infinite variety of structures. There are a variety of other components in the wall including protein, and lignin. Let's look at these wall components in more detail:

A. Cellulose: β1,4-glucan (*structure provided in class*). Made of as many as 25,000 individual glucose molecules. Every other molecule (called residues) is "upside down". Cellobiose (glucose-glucose disaccharide) is the basic building block. Cellulose readily forms hydrogen bonds with itself (intra-molecular H-bonds) and with other cellulose chains (inter-molecular H-bonds). A cellulose chain will form hydrogen bonds with about 36 other chains to yield a microfibril. This is somewhat analogous to the formation of a thick rope from thin fibers. Microfibrils are 5-12 nm wide and give the wall strength - they have a tensile strength equivalent to steel. Some regions of the microfibrils are highly crystalline while others are more "amorphous".

B. Cross-linking glycans (Hemicellulose): Diverse group of carbohydrates that used to be called hemicellulose. Characterized by being soluble in strong alkali. They are linear (straight), flat, with a β-1,4 backbone and relatively short side chains. Two common types include xyloglucans and glucuronarabinoxylans. Other less common ones include glucomannans, galactoglucomannans, and galactomannans. The main feature of this group is that they don't aggregate with themselves - in other words, they don't form microfibrils. However, they form hydrogen bonds with cellulose and hence the reason they are called "*cross-linking glycans*". There may be a fucose sugar at the end of the side chains which may help keep the molecules planar by interacting with other regions of the chain.

C. Pectic polysaccharides: These are extracted from the wall with hot water or dilute acid or calcium chelators (like EDTA). They are the easiest constituents to remove from the wall. They form gels (*i.e.,* used in jelly making). Another diverse group of polysaccharides that are particularly rich in galacturonic acid (galacturonans = pectic acids). Polymers of primarily β 1,4 galacturonans (=polygalacturonans) are called homogalacturons (HGA) and are particularly common. These are helical in shape. Divalent cations, like calcium, also form cross-linkages to join adjacent polymers creating a gel. Pectic polysaccharides can also be cross-linked by dihydrocinnamic or diferulic acids. The HGA's (galacturonans) are initially secreted from the golgi as methylated polymers; the methyl groups are removed by pectin methylesterase to initiate calcium binding.

Other pectic acids include Rhamnogalacturonan II (RGII) which features rhamnose and galacturonic acid in combination with a large diversity of other sugars in varying linkages. Dimers of RGII can be cross-linked by boron atoms linked to apiose sugars in a side chain.

Although most pectic polysaccharides are acidic, others are composed of neutral sugars including arabinans and galactans. The pectic polysaccharides serve a variety of functions including determining wall porosity, providing a charged wall surface for cell-cell adhesion (middle lamella), cell-cell recognition, pathogen recognition and others.

D. Protein: Wall proteins are typically glycoproteins (polypeptide backbone with carbohydrate sidechains). The proteins are particularly rich in the amino acids hydroxyproline (hydroxyproline-rich glycoprotein, HPRG), proline (proline-rich protein, PRP), and glycine (glycine-rich protein, GRP). These proteins form rods (HRGP, PRP) or beta-pleated sheets (GRP). Extensin is a well-studied HRGP. HRGP is induced by wounding and pathogen attack. The wall proteins also have a structural role since: (1) the amino acids are characteristic of other structural proteins such as collagen and gelatin; and (2) to extract the protein from the wall requires destructive conditions. Protein appears to be cross-linked to pectic substances and may have sites for lignification. The proteins may serve as the scaffolding used to construct the other wall components.

Another group of wall proteins are heavily glycosylated with arabinose and galactose. These arabinogalactan proteins, or AGP's, seem to be tissue specific and may function in cell signaling. They may be important in embryogenesis and growth and guidance of the pollen tube.

E. Lignin: Polymer of phenolics, especially phenylpropanoids. Lignin is primarily a strengthening agent in the wall. It also resists fungal/pathogen attack.

F. Suberin, wax, cutin: A variety of lipids are associated with the wall for strength and waterproofing.

G. Water: The wall is largely hydrated and comprised of between 75-80% water. This is responsible for some of the wall properties. For example, hydrated walls have greater flexibility and extensibility than non-hydrated walls.

II. MORPHOLOGY OF THE CELL WALL

There are three major regions of the wall:

1. Middle lamella - outermost layer, glue that binds adjacent cells, composed primarily of pectic polysaccharides.
2. Primary wall - wall deposited by cells before and during active growth. The primary wall of cultured sycamore cells is comprised of pectic polysaccharides (ca. 30%), cross-linking glycans (hemicellulose; ca 25%), cellulose (15-30%) and protein (ca. 20%) (see Darvill et al, 1980). The actual content of the wall components varies with species and age. All plant cells have a middle lamella and primary wall.
3. Secondary Wall - some cells deposit additional layers inside the primary wall. This occurs after growth stops or when the cells begins to differentiate (specialize). The secondary wall is mainly for support and is comprised primarily of cellulose and lignin. Often can distinguish distinct layers, S1, S2 and S3 - which differ in the orientation, or direction, of the cellulose microfibrils.

III. TIRE ANALOGY FOR THE CELL WALL

The wall is similar to a tire that has a series of steel belts or cords embedded in an amorphous matrix of rubber. In the plant cell wall, the "cords" are analogous to the cellulose microfibrils and they provide the structural strength of the wall. The matrix of the wall is analogous to the rubber in the tire and is comprised of non-cellulosic wall components. How are the various wall polymers assembled? It appears that:

1. cross-linking glycans (hemicellulosic polysaccharides) are hydrogen bonded to the cellulose microfibrils
2. cross-linking glycans may also be entrapped inside cellulose microfibrils as they form
3. the different types of pectic polysaccharides are covalently bonded to one another

4. calcium bridges link pectic acids
5. connections between the protein and other wall polymers are still not clear
6. pectic polysaccharides and cross-linking glycans interact
7. cross-linking glycans are linked by ferulic acid bridges or boron

IV. WALL FORMATION

The cell wall is made during cell division when the cell plate is formed between daughter cell nuclei. The cell plate forms from a series of vesicles produced by the golgi apparatus. The vesicles migrate along the cytoskeleton and move to the cell equator. The vesicles coalesce and dump their contents. The membranes of the vesicle become the new cell membrane. The golgi synthesizes the non-cellulosic polysaccharides. At first, the golgi vesicles contain mostly pectic polysaccharides that are used to build the middle lamella. As the wall is deposited, other non-cellulosic polysaccharides are made in the golgi and transported to the growing wall.

Cellulose is made at the cell surface. The process is catalyzed by the enzyme cellulose synthase that occurs in a rosette complex in the membrane. Cellulose synthase, which is initially made in by the ribosomes (rough ER) and move from the ER to vesicles to golgi to vesicle to cell membrane. The enzyme apparently has two catalytic sites that transfer two glucoses at a time (*i.e.*, cellobiose) from UDP-glucose to the growing cellulose chain. Sucrose may supply the glucose that binds to the UDP. Wall protein is presumably incorporated into the wall in a similar fashion.

Remember that the wall is made from the outside in. Thus, as the wall gets thicker the lumen (space within the wall) gets smaller.

Exactly how the wall components join together to form the wall once they are in place is not completely understood. Two methods seem likely:

1. self assembly. This means that the wall components spontaneously aggregate; and
2. enzymatic assembly – various enzymatic reactions (XET) are designed for wall assembly. For example, one group of enzymes "stitches" xylans together in the wall to form long chains. Oxidases may catalyze additional cross-linking between wall components and pectin methyl esterase may play an important role (see below).

V. STRONG WALL/CELL EXPANSION PARADOX

(don't you just love a good paradox?)

How can the wall be strong (it must withstand pressures of 100 MPa!), yet still allow for expansion? Good question, eh? The answer requires that the wall:

A. Be capable of expansion: In other words, only cells with primary walls are capable of growth since the formation of the secondary wall precludes further expansion of the cell. The sequence of microfibril orientation changes during development. Initially the microfibrils are laid down somewhat randomly (isotropically). Such a cell can expand in any direction. As the cell matures, most microfibrils are laid down laterally, like the hoops of a barrel, which restricts lateral growth but permits growth in length. As the cell elongates the microfibrils take on an overlapping cross-hatched pattern, similar to fiberglass. This occurs because the cell expands like a slinky - the width of the cell doesn't change by the microfibrils become aligned in the direction of growth just like the spring. This overlapping of microfibrils, which is strong and lightweight, prohibits further expansion.

But, what determines the orientation of the microfibrils? They are correlated with the direction of the microtubules in the cell. Evidence: treating a cell with colchicine or oryzalin (which inhibit microtubule formation) destroys the orientation of the microfibrils. The microtubules apparently direct the cellulose synthesizing enzymes to the plasma membrane.

In addition to cellulose microfibril orientation, mature walls apparently loose their ability to expand because the wall components become resistant to loosening-activities. This would occur if there were increased cross-linking between wall components during maturation. This would result from:

1. producing wall polysaccharides in a form that makes tighter complexes with cellulose or other materials
2. increasing the lignin in the wall would increase cross-links between polymers
3. de-esterifying the pectic acids would increase calcium bridges;

B. Loosening (or Stress relaxation) the wall at the appropriate time: Even though the microfibrils may be in the proper position to permit loosening, the wall is still rather strong. Recall that our wall model proposed strong and weak links between the wall components. When the wall is loosened bonds (*i.e.*, H-bonds) are temporarily broken to allow the wall components to slide or creep past one another. So, how is the wall temporarily loosened?

1. Protons are the primary wall loosening factor (**Acid Growth Hypothesis**). This idea was first proposed by David Rayle and R. Cleland in 1970. Some evidence:
   - acid buffers stimulate elongation and rapid responses 5-15 min even in non-living tissues (Evans,1974);
   - acid secretion is associated with sites of cell elongation (see Evans & Mulkey, 1981)
   - Fusicoccin, a diterpene glycoside extracted from a fungus, stimulates proton secretion (activates a H^+/K^+ pump) and stimulates elongation.
2. Mechanism of proton action: Protons stimulate wall loosening by:
   - disrupting acid-labile bonds such as H-bonds and calcium bridges; and
   - enhancing the activity of enzymes that break wall cross-links including H-bonds and calcium bridges. Evidence for the enzyme involvement includes: (1) when primary walls are heated or treated with protein denaturing agents they can't be "loosened" by acid; and (2) adding proteins extracted from growing walls to heat-treated walls restores the acid response.
   - Expansins – appear to be the primary wall-loosening enzymes. This class of proteins are activated by low pH and break the hydrogen bonds between cellulose and the cross-linking glycans. Other candidates for enzymes involved include: (1) pectin methyl esterase which would break the calcium bridges between pectins by esterifying the carboxyl groups; and (2) hydrolases – which would hydrolyze the cross-linking glycans (hemicelluloses). For example, xyloglucan endotransglycosylase (XET) has been shown to cleave cross-linking glycans that could allow slippage of the wall components
3. The acid effect is induced by indole-3-acetic acid (IAA, auxin), one of the major plant hormones. IAA stimulates proton excretion and cell growth/elongation. Evidence:
   - peeled coleoptiles + IAA - medium acidic; peeled coleoptiles + water - not acidic; and
   - flooding auxin-treated tissue with neutral buffers prevents the growth response.

4. Mechanism of Auxin Action - How does auxin stimulate proton excretion and wall elongation? There are two ideas:

Hypothesis 1: Auxin activates pre-existing H^+-ATPase pump proteins in the cell membrane. These proteins transport protons from the protoplast into the wall. Auxin probably first binds to a receptor molecule and this complex then actives the pump. This process is active - thus the pump requires ATP. Evidence: ATP stimulated acidification is observed soon after auxin treatment.

Hypothesis 2: Auxin stimulates transcription and translation. Transcription/translation (protein synthesis) would be required to produce proton pump proteins (a wonderful alliteration), respiratory enzymes to provide ATP to power the process; and even enzymes for the synthesis of wall components and cell solutes (see C. & D. below). Evidence for the involvement of transcription/translation:

- Nooden (1968) found that artichoke disks increased in size when incubated with IAA but that the addition of antimycin (a protein synthesis inhibitor) prevented this response;
- soybean hypocotyls incubated with 2,4-D (an analog of IAA) produce at least 3 new polypeptides within three hours (Zurfluh & Guilfoyle, 1980);
- *in vitro* translation of mRNA occurs within 15 minutes of IAA treatment
- The proton effect is short-lived. Cell elongation stops 30-60 minutes after acidification. Continuous elongation requires longer term metabolic changes such as protein synthesis.

C. Wall synthesis occurs: As the cell grows, wall synthesis needs to occur. Think about the color of a balloon as it is blown up - it gets lighter in color as the balloon gets larger because the thickness of the balloon decreases as it expands and stretches. Using this logic, we expect that plant cells should become thinner as they expand. Right? Wrong - cell walls remain a relatively uniform thickness throughout cell growth. Thus, we can conclude that new wall material must be made during cell elongation.

D. Enhanced solute synthesis: The solute concentration of the cell remains constant during cell enlargement. This suggests that solutes are being synthesized since the volume of the cell is increasing. Maintaining a high solute concentration is necessary to allow for water uptake.

E. Lock wall in place after expansion is complete: Once wall elongation is completed, the cell needs to "lock it" in place. This likely happens as the temporary bonds that were broken reform, and due to increased interactions (including enzymatic) between wall molecules.

F. Water Uptake/Pressure

VI. FUNCTIONS OF THE CELL WALL:

The cell wall serves a variety of purposes including:

1. Maintaining/determining cell shape (analogous to an external skeleton for every cell). Since protoplasts are invariably round, this is good evidence that the wall determines the shape of plant cells.
2. Support and mechanical strength (allows plants to get tall, hold out thin leaves to obtain light)
3. prevents the cell membrane from bursting in a hypotonic medium (*i.e.,* resists water pressure)
4. controls the rate and direction of cell growth and regulates cell volume
5. ultimately responsible for the plant architectural design and controlling plant morphogenesis since the wall dictates that plants develop by cell addition
6. has a metabolic role (*i.e.,* some of the proteins in the wall are enzymes for transport, secretion)
7. physical barrier to: (a) pathogens; and (b) water in suberized cells. However, remember that the wall is very porous and allows the free passage of small molecules, including proteins up to 60,000 MW. The pores are about 4 nm (Tepfert & Taylor 1987)
8. carbohydrate storage - the components of the wall can be reused in other metabolic processes (especially in seeds). Thus, in one sense the wall serves as a storage for carbohydrates
9. signaling - fragments of wall, called oligosaccharins, act as hormones. Oligosaccharins, which can result from normal development or pathogen attack, serve a variety of functions including: (a) stimulate ethylene synthesis; (b) induce phytoalexin synthesis; (c) induce chitinase and other enzymes; (d) increase cytoplasmic calcium levels and (d) cause an "oxidative burst". This burst produces hydrogen peroxide, superoxide and other active oxygen species that attack the pathogen directly or cause increased cross-links in the wall making the wall harder to penetrate.

Let's look at how this system works. Consider a pathogenic fungus like *Phytophthora*. In contact with the host plant the fungus releases enzymes such as pectinase that break down plant wall components into oligosaccharins. The oligosaccharins stimulate the oxidative burst and phytoalexin synthesis, both which will deter the advance of the fungus. In addition, the oligosaccharins stimulate chitinase and glucanase production in the plant. These are released and begin to digest the fungal wall. The fragments of fungal wall also act as oligosaccharins in the plant to further induce phytoalexin synthesis. Cool!

10. recognition responses - for example: (a) the wall of roots of legumes is important in the nitrogen-fixing bacteria colonizing the root to form nodules; and (b) pollen-style interactions at mediated by wall chemistry.

11. economic products - cell walls are important for products such as paper, wood, fiber, energy, shelter, and even roughage in our diet.

CHAPTER - 8

Biological Membrane and Function

COMMON FEATURES OF BIOLOGICAL MEMBRANES

- Membranes are sheetlike, just a few molecules thick and form closed boundaries between cell compartments.
- Membranes contain lipids and proteins, with small amounts of carbohydrayes linked to the lipids and proteins.
- Lipids in membranes are small with hydrophobic and hydrophilic portions. Lipid bilayers provide a barrier to the diffusion of polar molecules.
- Characteristic functions of membranes are mediated by specific proteins, serving as pumps, channels, receptors, energy transducers and enzymes.
- Membrane components associate through noncovalent interactions.
- Membranes are asymmetrical, with two sides of the membrane differing from each other.
- Lipid and protein molecules often diffuse rapidly in the plane of the membrane.
- Central transport of ions and molecules into and out of the cell.
- Generate proton gradients for ATP production by oxidative phosphorylation.
- Receptors bind extracellular signals and transduce the signal to cell interior.

- Structure:
  - Glycerophospholipids and glycosphingolipids form bilayers.
  - Noncovalent interactions hold lipids together.
  - 5-6 nm thick and made of 2 leaflets to form a lipid bilayer driven by hydrophobic effects.
  - About 40% lipid and 50% proteins by mass, with about 10% carbohydrates.
- Protein and lipid composition varies among membranes but all have same basic structure ® Singer and Nicholson fluid mosaic model in 1972.

BILAYER FORMATION

Hydrophobic interactions provide the primary driving force for the formation of bilayers.

Simple Bilayer

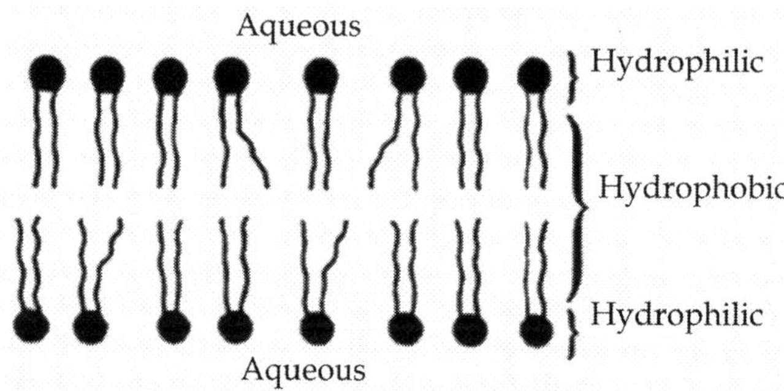

Saturated fatty acid chains **pack easily** and have a **higher melting temperature (Tm)**. Butter is a solid at room temperature so has a high Tm.

Unsaturated fatty acids have a **lower Tm**. Canola oil is a liquid at room temperature so has a low Tm.

Cholesterol **impedes motion** of the hydrocarbon tails making membranes **less fluid**.

The degree of saturation of the "tails" affects the stability of the membrane. Saturated fats pack more easily than unsaturated fats. A high percentage of unsaturated fats lowers the temperature at which a membrane will become rigid.

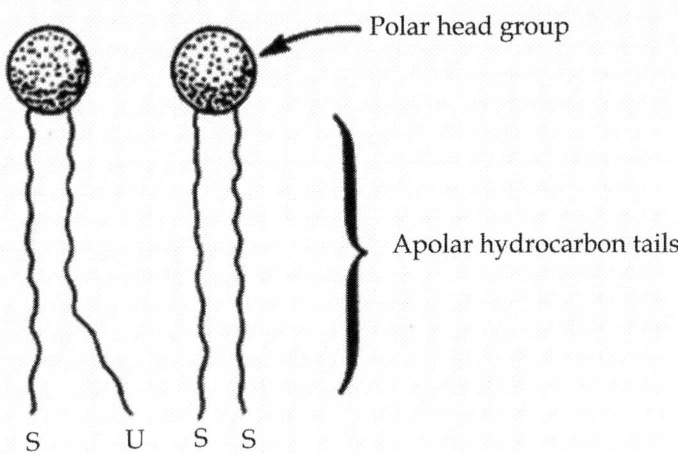

Fluid Mosaic Model

- The lipids are arranged in a bilayer, which is both a permeability barrier and solvent for integral proteins.
- Some lipids interact with specific proteins to produce characteristic functions of the membrane.
- Lipids diffuse laterally (horizontally) rapidly but transversely (vertically) slowly. Proteins diffuse laterally.

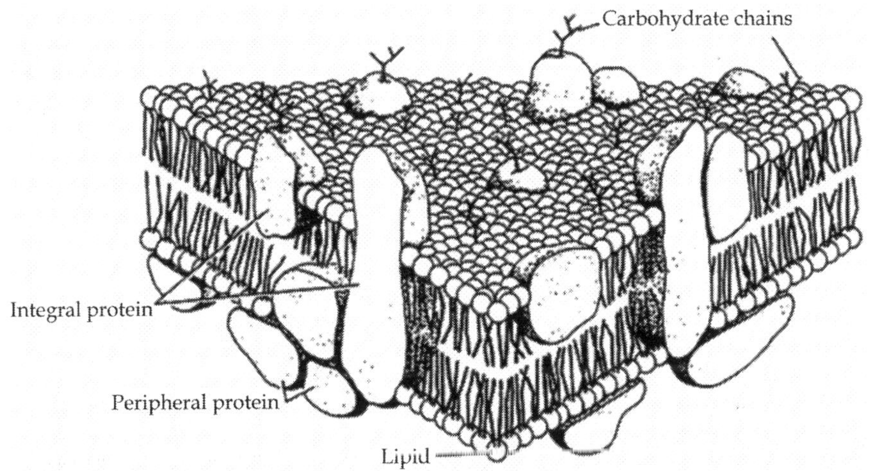

Membrane Carbohydrates

- Common monosaccharides associated with membranes are: glucose, galactose, fructose and mannose.
- Polysaccharides attached to membranes are built from monosaccharides attached to each other via glycosidic bonds.
- Hexoses can be linked from any of 5 positions on the molecule → possibility of numerous different structures, i.e. "*high information content*".
- Both membrane proteins and lipids can be glycosylated.
- In glycoproteins, the sugar residues are attached to nitrogen of Asn (N-linked) or via hydroxyl of a Ser or Thr (O-linked).
- Both glycoproteins and glycolipids are important in cell-cell recognition.

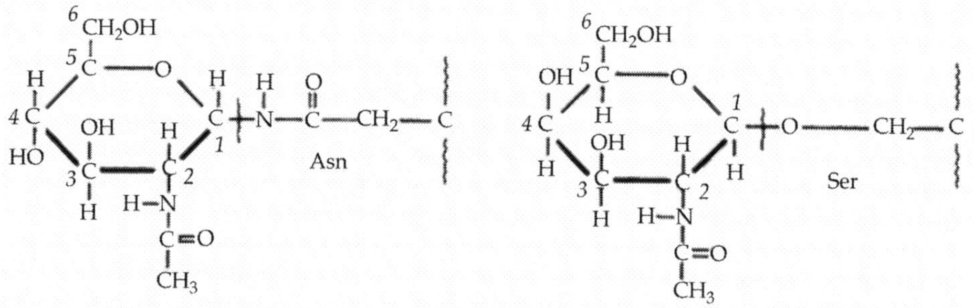

Membrane Asymmetry

1. Membrane components are asymmetrically distributed across the bilayer.
2. Membranes are asymmetrically oriented: pumps *drive transport* in one direction receptors *bind molecules* on the outside
3. Carbohysrates are asymmetric because they are found on the outside of the surface of the cell.

Sphingolipids and Glycolipids

Sphingomyelin (an important *spingolipid*)-

A lipid found in brain, blood cells and lung surfactant.

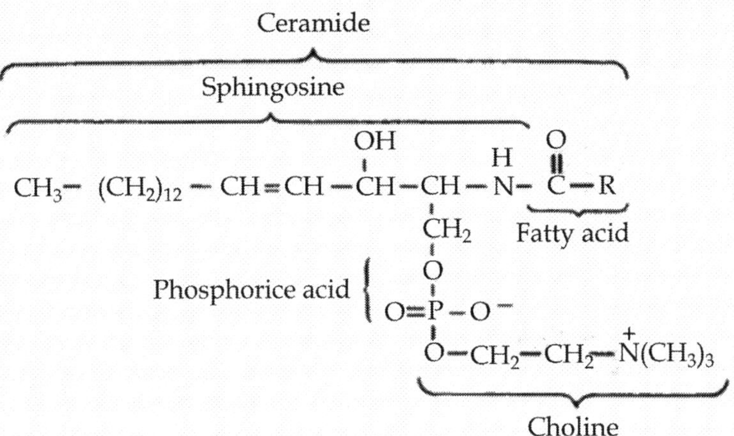

Glycolipids

- have an identical structure, except the phosphoryl choline headgroup is replaced with a monosachharide (sugar).

Peripheral Membrane Proteins

- Are either extracellular or intracellular.
- Though associated with membranes they can be easily removed.

Extracellular Proteins

- Protein ligands specific for cell surface receptors.
- Proteins of the extracellular matrix.

Intracellular Proteins

- Attach to lipid anchors or to integral proteins.
- These proteins can be modified covalently by lipids → localizes them to specific membranes.
- Many of these modified proteins are oncoproteins → cause cancer if they become malfunctional.

Integral Membrane Proteins

- Span the bilayer and thus have both intra- and extracellular domains.
- The transmembrane domain is always a-helical.

- They can not be removed from the membrane without very harsh treatments.

They fall into three classes: **antigens, receptors** and **translocators**.

1. **Antigens**:
    - Integral membrane proteins that are recognized by antibodies.
2. **Receptors**:
    - Are required for the specific action of hormones, transmitters and growth factors.
    - Three general classes of membranes receptors: growth factor receptor tyrosine kinases (RTKs), small molecule 7 transmembrane helix receptors, receptor channels.

(a) RTKs bind hormones such as insulin

They contain:

- an extracellular ligand binding domain
- a single a-helical transmembrane domain
- an intracellular domain which can phosphorylate tyrosine residues on proteins.

(b) Small Molecule 7 Transmembrane Helix Receptors

- bind hormones such as epinephrine and glucagon

Membrane fluidity

- Lipids can undergo lateral diffusion
- Can undergo transverse diffusion (one leaflet to another) but very rare.
- Membrane has an asymmetrical lipid distribution that is maintained by flippases or translocases that are ATP-driven.
- In 1970, Frye and Edidin demonstrated that proteins are also capable of diffusion by using heterocaryons, but occurs at a rate that is 100-500 times slower than lipids.
- Most membrane protein diffusion is limited by aggregation or attachment to cytoskeleton.
- Can examine distribution of membrane proteins by freeze-fracture electron microscopy.

- Membrane fluidity is dependent upon the flexibility of fatty acyl chains.
- Fully extended saturated fatty acyl chains show maximum van der Waals interactions.
- When heated, the chains become disordered → less interactions → membrane "shrinks" in size due to less extension of tails → due to rotation around C-C bond.
- For lipids with unsaturated acyl chains, kink disrupts ordered packing and increases membrane fluidity → decreases phase transition temperature (becomes more fluid at lower temperature).
- Some organisms can alter their membrane fluidity by adjusting the ratio of unsaturated to saturated fatty acids.

  e.g. bacteria grown at low temperature increase the proportion of unsaturated fatty acyl groups.

  e.g. warm-blooded animals have less variability in that ratio because of the lack of temperature fluctuations.

  exception: reindeer leg has increased number of fatty acyl groups as get closer to hoof → membrane can remain more fluid at lower temperatures.
- Cholesterol also affects membrane fluidity.
  - Accounts for 20-25% of lipid mass of membrane.
  - Broadens the phase-transition temperature.
  - Intercalation of cholesterol between membrane lipids restricts mobility of fatty acyl chains → fluidity decreases.
  - Helps maintain constant membrane fluidity despite changes in temperature and degree of fatty acid saturation.

The Cell Membrane Forms a Barrier Between the Cell and the Outside World

A membrane around the cell is necessary for many reasons:

Cell must be able to retain the molecules it makes by metabolism

Must be able to exclude unwanted molecules

Membranes separate compartments with differenct chemical reactions

Excitable phenomena. such as nerve conduction, use membranes

Cell recognition and attachment involves membranes

Many metabolic functions, such as oxidative phosphorylation, are organized around membranes

Signal systems for cell regulation are based upon membranes

Cell Membranes are Primarily Phospholipid Bilayers

Cell membranes are phospholipid bilayers (2 molecular layers)

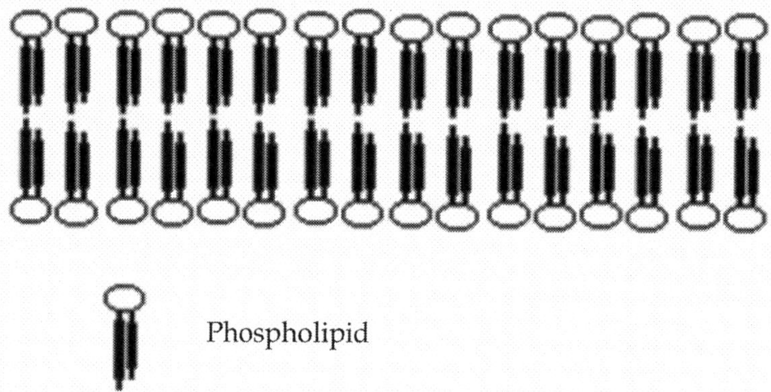

Phospholipids are amphiphilic ("with 2 loves"): head wants to be in water, tail wants to be in lipid (hydrophobic)

Form a bilayer with the tails together

Phospholipids = glycerol + 2 fatty acids + phosphate + polar molecule **X** (i.e., choline)

```
        O
        ||     H
        ||     |
X-O-P-O-C-H
        |      |
        O      |
               |   O
               |    \\  H H H H HH HH H H HH H H H
          H-C-O-C-C-C-C-C-C-C-C-C-C-C-C-C-C-C-H
               |        H H H H H H HH H H H H H HH
               |
               |   O
               |    \\  H H H H HH HH H H HH H H H
          H-C-O-C-C-C-C-C-C-C-C-C-C-C-C-C-C-C-H
               |        H H H H H H HH H H H H H HH
               H
```

Fatty acid tails are hydrophobic; phosphate and polar molecule are hydrophilic (rotate to opposite side of molecule)

Different types of polar groups (**X**): choline, ethanolamine, inositol, serine, etc

If the X group is choline the phospholipid is called phosphatidylcholine (PC); the other X groups give phosphatidylethanolamine (PE), phosphatidylinositol (PI) and phosphatidylserine (PS)

Also many different types of fatty acids, some saturated (no double bonds), others unsaturated (1 or more double bonds)

Many different types of fatty acids used, most 16 to 18 carbons long

Unsaturated fatty acids don't pack well -> membrane is more fluid

Cholesterol (another lipid) stabilizes cell membranes; fits between phospholipid tails

Phospholipids Form Structures that Keep their Tails out of Water

Monolayers formed at air/water surface

Tails in air, heads in water

Studied by Benjamin Franklin (1770): observed that oil thrown overboard calmed the sea

Monolayers reduce the surface tension of water

Found that approx. 1 teaspoonful of oil would spread out to cover about a half acre of water

If he had divided the 2 figures he would have had the 1st estimate of molecular sizes:

thickness = ("not more than a teaspoonful")/("perhaps half an acre")

using modern units, this calculation would have given an oil layer thickness of about 10 angstroms, close to the true value

Micelles: with oil droplets phospholipids accumulate at the surface with their tails in the oil

This keeps oil droplets in suspension (detergent effect)

Lipids in the blood are in the form of micelles (chylomicrons)

Liposomes are formed when a mixture of phopholipids and water is shaken or sonicated

Tails are together in the center with heads on both surfaces

An artificial bilayer membrane

Liposomes have medical uses: delivering drugs, gene therapy, etc.

Membranes Also Contain Proteins

Proteins that penetrate the membrane (intrinsic proteins) have hydrophobic sections ~25 amino acids long (example: glucose transporter)

The bilayer penetrated by proteins is referred to as the fluid mosaic model of the membrane

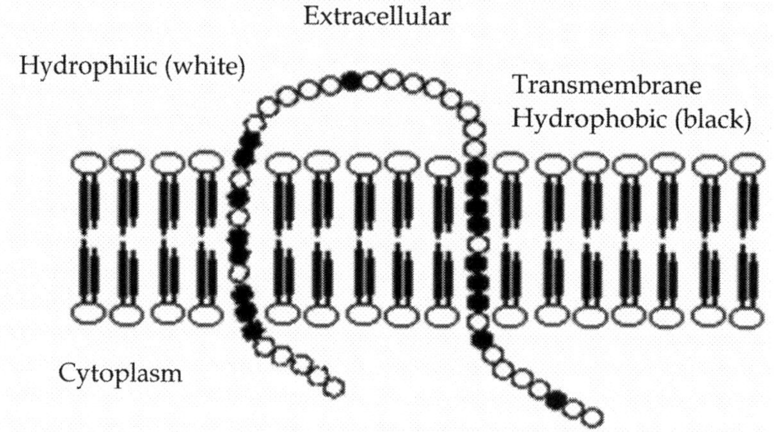

Membrane proteins have many functions:

receptors for hormones

pumps for transporting materials across the membrane

ion channels

adhesion molecules for holding cells to extracellular matrix

cell recognition antigens

Carbohydrates May be Attached to the Outer Membrane Surface

Carbohydrate chains attach to some amino acid side groups of membrane proteins: usually to serine, threonine or asparagine

Usually small branched oligosaccharides; an oligosaccharide is a small polysaccharide

Proteins with sugar groups attached are called glycoproteins

Carbohydrates added to proteins in the endoplasmic reticulum & Golgi apparatus

Some sugars are also attached to lipids, forming glycolipids

Carbohydrates found only on outer (extracellular) side of cell membrane (also found inside lysosomes and other vesicles)- none exposed to cytosol

Human ABO blood group antigens are oligosaccharides

The Cell Membrane is Asymmetric

As noted, carbohydrates are found only on the outer side of the bilayer

The different types of phospholipids are distributed asymmetrically in the 2 phospholipid layers: PC is usually in the outer layer and PS, PI and PE are in the inner bilayer.

The asymmetry is maintained by a "lipid pump" which requires ATP energy

Purpose of the asymmetry is unknown, but may have something to do with membrane fusion and bending; membranes bend during cell movement, endocytosis, etc. Some membranes fuse readily, while others do not.

The inner surface is supported by the cytoskeleton; some membrane proteins attach to the cytoskeleton

Membrane proteins are oriented

The Cell Membrane is in a Fluid State

The bilayer is like a 2-dimensional fluid- there is not much movement across the membrane, but there is a lot of sideways (lateral) movement

Fluidity depends upon types of fatty acids in phospholipids:

unsaturated fatty acids have kinks

don't pack as well → more fluid

Fluidity is important for function; if membrane becomes more solid transport will be slowed

Some organisms increase unsaturated fatty acids in membranes when exposed to cold weather

Fluidity also important for fusion of membranes and sealing of leaks

Membranes Fuse and Seal Leaks

If a cell is cut in half or poked with a small probe it will usually seal the wound with little damage

Fusion of membranes is an important biological process:

Fusion of sperm and egg

Muscle cells fuse in development to produce a single giant cell

Inside the cell vesicles fuse together

Example 1: vesicles from endoplasmic reticulum fuse with the Golgi apparatus

Example 2: new membrane material made by smooth ER must fuse with the cell membrane

Example 3: glucose transport proteins stored in vesicles in muscle and fat tissue; vesicles fuse with cell membrane under influence of insulin

Membranes must also fuse when vesicles pinch off from cell membrane, in endocytosis and phagocytosis

Artificial membrane fusion important in biology and medicine

Delivery of drugs with liposomes

Hybridomas: fusion of 2 cells to produce antibodies

Functions of Cell Membrane

Some of the important functions of the cell membrane are as follows :

(i) It acts as a permeability barrier which controls and co-ordinates the rate of substrate transfer and diffusion. Compounds with polar groups such as -OH, _COOH, NH_2, -CHO and inorganic salts enter the cell slowly. Non-polar compounds, alcohol, chloform etc. penetrate rapidly. Cell membrane is impermeable to polysaccharides, phospholipids and proteins.

(ii) It acts as a cytoskeleton providing mechanical frames on which enzymes can be specifically oriented.

(iii) It acts as a vehicle for transport of substances from one organelle to another, from inside to the outside of the cell and from outside to the inside of the cell.

(iv) It acts as element, supporting the synthesis of various macro-molecules, particulary in chloroplast, mitochondia and cell membrane of prokaryotes.

Fluid mosaic : functions of the membrane

The plasma membrane performs various functions like: Selective permeability: Membranes allow only certain groups of molecules to pass through,hydrophobic interior produces impermeability while specific transport proteins allows permeability of hydrophilic molecules membrane proteinsh..

Functions of semi permeable membrane

To maintain cell shape with help of cytoskeleton. Regulation of cell potential. Transport of selected molecules or ions by diffusion (simple and facilitated) or active transport. Cell-cell signaling. Helps in immune response. Artificial semi permeable membrane is used in dialysis treatment for patients with kidney failure.

Function of the Nuclear membrane

The nuclear membrane surrounds the nucleus that is covered with pores and it controls nuclear traffic and the nuclear membrane holds the nucleus together. The nuclear membrane encloses the nucleus of the cell that controls the things which enters and leaves the nucleus. So it is also called as the nuclear enve..

Functions of golgi apparatus membrane

The main function of golgi apparatus is to as an area for storage, processing and packaging of various cellular secretions. It packages materials synthesized in the cell and dispatches them either to intracellular targets like plasma membrane and lysosomes or extracellular targets. It produces vacuoles or secretory ves..

Plasma membrane function

Function of plasma membrane is important in survival of all living cells. It not only protects the interior part of cell also helps the cells to maintain its shape. Main function of plasma membrane is as follows: Protect and separate the interior part of cell (protoplasm) from external environment..

Function of lipid membrane

The lipid bilayer around the cell separates it from surroundings and membrane around cell organelles form compartments within the cell that restricts the permeability of different molecules and helps in proper function of cell. A wide variety of molecules act as ligands, which carry the signal through receptors.

CHAPTER - 9

Membrane Transport

ALL MOLECULES MOVE CONTINUOUSLY BY SIMPLE DIFFUSION

Heat energy causes molecules to move randomly (sometimes called Brownian motion)

If the concentration of molecules is different in 2 regions (this produces a concentration gradient), diffusion will cause molecules to move from a region of high concentration to one of low concentration

The higher the concentration gradient the more rapid the net diffusion

Diffusion evens out the concentrations so they are equal everywhere (maximum entropy)

If 2 different substances have the same concentration gradient, usually one will move faster than the other

Differences in speed between molecules with the same concentration gradient are given in terms of diffusion constants

A substance with a high diffusion constant moves faster than one with a low diffusion constant

In general, large molecules move slower than small molecules

Diffusion across a membrane is called permeability, and membrane diffusion constants are called permeability constants

Membrane Fluxes are Driven by Forces Existing Across the Cell Membrane

A flow or flux of a material is proportional to the gradient of force acting on the molecule. The forces arise because conditions are different on the 2 sides of the membrane:

Concentration differences

Voltage differences

Pressure differences

Osmotic pressure differences

Typical forces and fluxes are (the forces are the differences in concentration or other conditions, divided by the membrane thickness).

Type of Flux	Conditions Different	Force
Diffusional Flux	Concentrations, C	Concentration gradient
Electrical Current	Voltage, E	Voltage gradient
Bulk (Volume) Flow	Pressure, P	Pressure gradient
Osmosis (Volume) Flow	Osmotic Pressure, OP	Osmotic pressure gradient

More than one force may act upon a molecule. For example, an ion will be driven by both voltage and concentration gradients. Water flow typically is affected by both pressure and osmotic pressure gradients. Since very large amounts of water flow in osmosis there will be significant changes in the volume of the cell.

Simple Diffusion Across Membranes is Called Permeability

The flow or flux of materials across membranes by simple diffusion is called the permeability

Simple diffusion spontaneous; it does not require energy from ATP

In simple diffusion the flux is proportional to the concentration gradient; no saturation is seen

The permeability coefficient is the diffusion constant in the membrane divided by the membrane thickness; substances penetrating the membrane rapidly have high permeability constant

Flux = (Permeability Constant) x (Concentration Difference)

The flux is always downhill, from a high concentration to a lower one

Hydrophobic Substances Have a High Permeability Through Bilayer Membranes

Hydrophobic chemicals cross membranes faster than chemicals that like water

First stage of flow across a membrane is partition from the water into the membrane; hydrophobic molecules partition better into the lipid membrane

Hydrophobic substances have high permeability coefficients

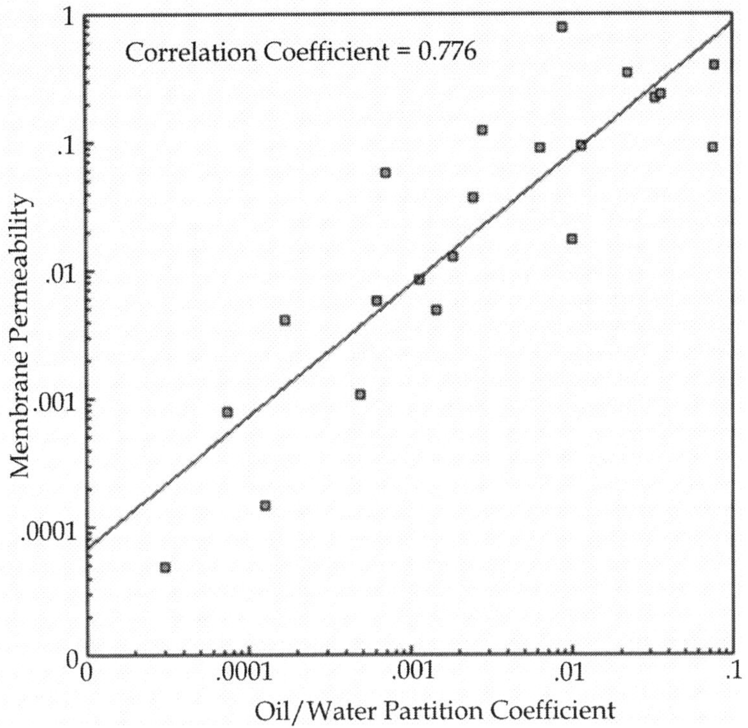

Many biological chemicals are deliberately made hydrophobic to increase their rate of penetration into cells.

Examples: many drugs, pesticides such as DDT

Hydrophobicity measured by oil/water partition

Osmosis Moves Water Across Biological Membranes and Causes Volume Changes

Osmosis = movement of water from low osmotic pressure (dilute solution) to high osmotic pressure (concentrated solution)

Low Osmotic Pressure

High Osmotic Pressure

Water Flow

High Water Concentration

Low Water Concentration

It is useful to think of a dilute solution as having a high water concentration and a concentrated solution as having a lower water concentration. Then the water flow goes from high water to low water concentration. A more accurate picture is that solutes lower the free energy of water.

In a mixture the substance present in the greatest amount is called the solvent; in biology the solvent is almost always water

Solutes are the dissolved substances in the solvent.

Osmotic pressure is computed by adding up the total concentration of solutes (ions are counted separately)

Solution with low solute concentration has low osmotic pressure; the water has high free energy

Solution with high solute concentration has high osmotic pressure; the water has low free energy

Osmosis is a type of bulk flow; it is not the same as simple diffusion

Osmosis is passive: doesn't require ATP energy

Cells Swell in Hypotonic Solutions and Shrink in Hypertonic Ones

All cells have osmotic problems, especially those which live in dilute solutions (such as freshwater)

There is so much water that osmosis often causes significant volume changes, causing swelling or shrinking

Biologists sometimes talk of osmosis in terms of tonicity:

If the external solution balances the osmotic pressure of the cytoplasm it is said to be isotonic.

If the external solution is more dilute than the cytoplasm it is hypotonic

If the external solution is more concentrated it is hypertonic.

Cell B is in an isotonic solution. What kind of solutions are cells A and C in?

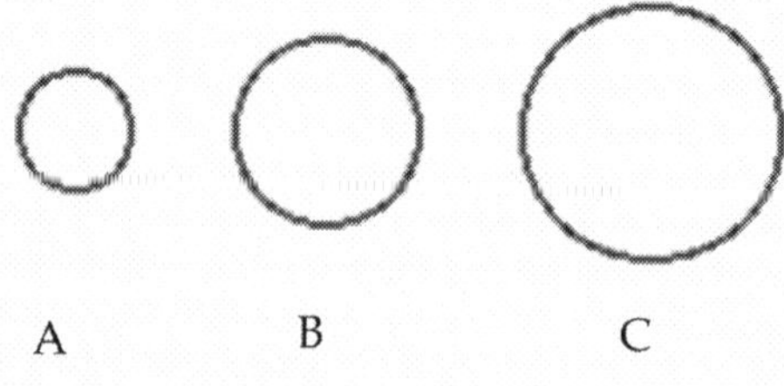

Cells Have Developed Different Ways of Combating Osmosis

Osmotic swelling dilutes the cytosol and can eventually cause the cell to burst

Cells have different ways of preventing excessive swelling:

Cell walls of plant, fungal and bacterial cells are rigid and prevent swelling- the walls are strong enough to allow a fairly high pressure gradient

Some protozoa have contractile vacuoles which store excess water and then squirt it out

Most cells pump ions out of the cell, which reduces the internal osmotic pressure; the most important pump is the Na/K pump

In Facilitated Diffusion Special Proteins Help Move Substances Across Membranes

Protein transport molecules are used to carry many substances across membranes

Very specific: allows cell to select substances taken up

Sensitive to inhibitors that react with protein side chains

ATP energy not required

Transport rate reaches a maximum when all of the protein transporters are being used

Cannot transport molecules against a concentration gradient (no "uphill" transport)

Example: glucose transporters

Glucose is hydrophilic, has very low permeability across lipid bilayer

7 proteins transporting glucose across cell membranes are known

Each has 12 hydrophobic sections imbedded in the membrane

One of the glucose transporters (GLUT 4) is controlled by insulin in muscle and fat tissue; insulin causes more glucose transporters to be inserted into the cell membrane- this helps to lower blood glucose

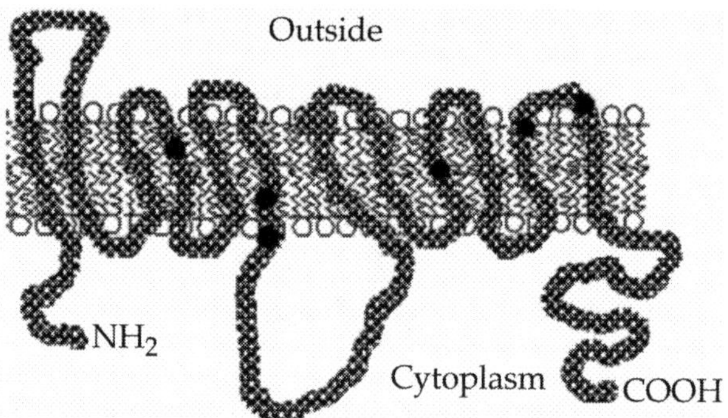

Some transport proteins form channels with "gates"; gates normally closed but open in response to electrical or chemical stimulus

Na and K channels of nerves- responsible for nerve action potentials

Some transporters carry more than 1 type of molecule (coupled transport)

Example: some sugar and amino acid transporters carry Na ion in addition to the organic molecule; both must be present

Endocytosis Can Bring Macromolecules Into the Cell

In endocytosis the cell membrane bends inward (invaginates), forming a vesicle (endosome) containing extracellular fluid and other substances

Can bring in large molecules such as proteins; such molecules would normally not diffuse across cell membranes

Some endocytic vesicles are coated with receptor molecules that selectively bind molecules

Example: LDL receptors (LDL = low density lipoprotein)

The macromolecules are usually digested by lysosomes

Active Transport Uses Energy to Pump Molecules Against a Concentration Gradient

Membrane pumps are proteins that use ATP energy to move substances across the cell membrane

Like facilitated diffusion the transport rate has a maximum which occurs when all of the transport molecules are being used

Also sensitive to inhibition by protein reagents

Can pump substances from a low concentration to a high concentration ("uphill" transport): called active transport

Example: the Na/K pump

Found in all cells

In humans may account for as much as 30% of basal metabolism

Pump is an ATPase because it splits ATP in its operation

Pumps 3 Na ions out of cell and 2 K ions in with each pump cycle:

3 Nas bind to sites exposed on inside of cell

ATP binds and is hydrolyzed to ADP, leaving a phosphate bound to the pump

Pump changes shape, exposing sites to outside of cell

3 Nas leave pump and 2 Ks are bound (to different sites)

Phosphate is split from pump

Shape changes again, exposing sites to inside

2Ks leave pump to inside

Inhibited by drugs such as digitalis, ouabain

Purposes:

Helps keep osmotic pressure of cell low (protects against swelling)

Charges the membrane electrically

Sodium gradient used for secondary active transport of amino acids and sugars

Active Transport Produces Concentration Gradients Across Membranes

Because active transport can pump uphill using ATP for an energy source it produces concentration gradients

Example 1: Na/K pump:

Reduces Na inside cell and raises K

Typically K inside the cell will be ~140 mM and outside it will be only ~5 mM (humans)

Na will be ~150 mM outside and ~10 mM inside the cell

Example 2: Ca pump:

Ca outside the cell is usually around 1 to 2 millimoles/liter

Inside the cell Ca is kept more than 1000 times lower by a Ca pump

Important because Ca within cell is very toxic if it gets too high

Concentration gradients represent stored energy- can be tapped for doing work

Concentration Gradients of Ions Will Electrically Charge Membranes

Voltage differences are caused by separation of charges

Ion pumps can charge cell membranes in 2 ways:

- They may pump more ions in one direction than the other
- They produce ion gradients- as ions diffuse back through the membrane they carry charge. One side will end up negative and the other positive
- Example: the Na/K pump sets up a gradient of K ions with a high concentration inside the cell
    - Some K will diffuse back out of the cell, carrying + charge
    - The outside of the cell will become positive and the inside negative
    - Most cells are charged with the inside negative due to the K gradient
    - Diffusion of Na would tend to make the inside positive, but Na is less important because it has a much lower permeability
    - Ion gradients are stored electrical energy, like batteries

Many Molecules Enter Cells by Secondary Active Transport

The Na concentration gradient is used to produce secondary active transport of sugars and amino acids

Some sugar and amino acid transporters must bind Na as well as the sugar or amino acid (coupled transport)

Both Na and the organic molecule must be present at the same time and on the same side of the membrane

Since there is more Na outside the cell, sugars and amino acids get transported mainly from the outside to the inside

The sugar and amino acid transporters do not use ATP directly, but ATP is required to set up the Na gradient

Knowledge of Transport Mechanisms is Important in Medicine

Many drugs are made to be hydrophobic so that they cross membranes more easily

Example: general anesthetics

Except for blood flow almost all water movement in body is osmosis

Kidney is an osmotic machine: adjusts body water volume by osmosis

Medical problems involving osmosis: pulmonary edema, childhood diarrhea, cholera, inflammation of tissues

Partial inhibition of the Na/K pump with cardiac glycosides strengthens the heartbeat

Mechanism:

inhibition of pump → more Na inside cell → exchange with Ca → more Ca inside heart cell → stronger contraction

Cholera treatment by oral rehydration therapy (ORT)

Cholera patients loose enormous amounts of water by osmotic diarrhea

They die because of dehydration

Cholera toxin causes secretion of Cl into intestine → osmosis & fluid loss

To treat this you should give fluids containing both sugar and NaCl

This causes transport of both sugar and NaCl into the blood (coupled transport)

Blood osmotic pressure rises causing osmotic flow of fluid back into the blood, reversing the loss

ORT can save as many as 95% of cholera victims

Comparison of Simple Diffusion, Facilitated Transport & Active Transport

Property	Simple Diffusion	Facilitated Transport	Active Transport
Requires special membrane proteins	No	Yes	Yes
Highly selective	No	Yes	Yes
Transport saturates	No	Yes	Yes
Can be inhibited	No	Yes	Yes
Uphill transport	No	No	Yes
Requires ATP energy	No	No	Yes

Note that most of the special properties of facilitated and active transport (those checked "yes") are due to the protein nature of the transport molecules.

CHAPTER - 10

Metabolism of Macromolecules

Metabolism - sum total of all chemical reactions in living cells

Catabolic reactions - degrade macromolecules and other molecules to release energy

Anabolic reactions - used to synthesize macromolecules for cell growth, repair, and reproduction

Can divide metabolism into 4 groups: carbohydrates, lipids, amino acids, nucleotides.

- within each group are a set of pathways
- arbitrarily set start and end points for ease of learning and reference
- pathways can take different forms:
    1) linear - product of one reaction is substrate for another

       e.g. glycolysis
    2) cyclic - regeneration of intermediates

       e.g. Krebs cycle
    3) spiral - same set of enzymes is used repeatedly

       e.g. fatty acid synthesis, β-oxidation
- each pathway may have branch points for metabolites to enter or leave

Why have metabolic reactions with so many steps?

1) energy input and output can be controlled
   - energy transfer occurs in discrete steps as it it transferred to acceptors a little at a time
2) enzymes can catalyze only a single step of a pathway
3) provides opportunities to establish control points, which are essential for cell function

METHODS OF METABOLIC PATHWAY REGULATION

1) **Feed back inhibition**
   - product of pathway controls its own rate of synthesis
   - occurs in the first committed step

     E_1 E_2 E_3
     A ® B ® C ® D

   - advantage is obvious ® prevention of intermediate accumulation
2) **Feed forward activation (positive feedback)**
   - metabolite produced early in pathway activates an enzyme later in pathway
   - also prevents accumulation of intermediates

     E_1 E_2 E_3 E_4
     A ® B ® C ® D ® E

3) **Allosteric activators and inhibitors**
4) **Covalent modification**
   - addition of phosphoryl groups via protein kinases
   - removal of phosphoryl groups via phosphatases

Major Catabolic Pathways

- begins with extracellular digestion of polymers (exogenous)
- amylase in mouth and intestine work on starch
- protein digestion starts in stomach and finished via pancreatic proteases and intestinal peptidases
- lipid digestion - triacylglycerols hydrolyzed to fatty acids by phospholipases
- absorption occurs in intestine ↔ blood → body

- can also have endogenous sources, such as glycogen and triacylglycerols
- catabolism yields 3 possible compounds:
  1) acetyl CoA
  2) nucleoside triphosphates
  3) reduced coenzymes
- starts with glycolysis (glucose catabolism), citric acid cycle, polysaccharide mobilization, oxidative phosphorylation
- nucleotides are metabolized for excretion, not energy production

Thermodynamics and Metabolism

- used to understand equilibrium and flux (flow of material through a metabolic pathway) in metabolism
- metabolic pathways are not at equilibrium, but at steady state (e.g. leaky bucket)
- free energy change (DG) is a measure of energy available to proceed in a chemical reaction

$$DG = G_{products} - G_{reactants}$$

- at equilibrium, DG = 0, no free energy available
- would like DG to be as small as possible (i.e. negative)
- Free energy change of a chemical reaction is expressed in terms of changes in heat content (enthalpy) and randomness (entropy)

D = LH - TLS

DH = change in enthalpy

T = temperature in ° Kelvin

DS = change in entropy

- when DG = -, reaction is spontaneous, and no energy input is needed
- when DG = +, must supply energy from outside
- when DG = 0, reaction is at equilibrium

DG°′ = standard free energy change of a biochemical reaction at standard conditions (pH 7.0; 25°C; 1M concentration of solute)

- DG°′ of a reaction is related to K_{eq} (equilibrium constant of a reaction)

$$A + B \rightarrow C + D$$

$$DG_{rxn} - (G_C + G_D) - (G_A - G_B)$$

$$Keq = \frac{[C][D]}{[A][B]}$$

$LG^{o\prime} = -2.303\ RTlog\ K_{eq}$ or $DG^{o\prime} = -RT\ ln\ K_{eq}$

R = gas constant 8.315 $JK^{-1}mol^{-1}$

- under ideal conditions (standard conditions):
- if Keq > 1, $DG^{o\prime}$ is negative and reaction will proceed to equilibrium
- if Keq = 1, $DG^{o\prime}$ =0 and reaction is at equilibrium
- if Keq <1, $DG^{o\prime}$ is positive
- DG and $DG^{o\prime}$ are related by the following equation:

$$LG = DG^{o\prime} + RT\ ln\ Q \qquad Q = \frac{[C][D]}{[A][B]}$$

R= 8.315$JK^{-1}mol^{-1}$

T = 298°K (25°C)

- free energy change is a measure of how far from equilibrium the system is poised
- DG, not $DG^{o\prime}$ determines spontaneity of a reaction and its direction
- means that some reactions have a -DG even if under standard conditions they have a +$DG^{o\prime}$.
- happens if Q is small or [A][B] >>[C][D]

Still find reactions that have a $DG^{o\prime}$ that is positive and still part of a metabolic pathway

How can these reactions with $DG^{o\prime}$ be made to go forward?

1) have other than "standard" concentrations
2) thermodynamic coupling

$DG^{o\prime}$ can be summed for a series of reactions

e.g. want A ® C

A ® B + C	$DG^{o\prime}$ = +5 kcal/mol
B ® D	$DG^{o\prime}$ = -8 kcal/mol
A ® D + C	$DG^{o\prime}$ = -3 kcal/mol

To make C from A is thermodynamically unfavorable, but if coupled to B, then becomes favorable.

3) couple reaction to ATP hydrolysis

- very common

  ATP ® ADP + P_i DGº′= -30 $kJmol^{-1}$

  ATP ® AMP + PP_i DGº′= -32$kJmol^{-1}$

  AMP ® adenosine + P_i DGº′ = -14 $kJmol^{-1}$

- also works with other nucleoside triphosphates, such as UTP, GTP, and GTP
- ADP and AMP are often allosteric modulators of some catabolic reactions
- ATP not effective in the role of allosteric modulator because its concentration is kept relatively constant in the cell
- cells typically maintain [ATP] of 2-10 mM, [ADP] <1 mM, and [AMP] <<1 mM
- the metabolic role of ATP to above problem:

  A ® B + C DGº′ = +5 kcal/mol

  ATP ® ADP + P_i DGº′ = -7.3 kcal/mol

  A + ATP– – –> B + C + ADP +P*i* DGº′ = -2.3 kcal/mol

There are several types of group transfer reactions that involve ATP:

1) Phosphoryl group transfer

- some metabolites have high phosphoryl group transfer potential (ability to transfer phosphoryl groups)

  - e.g. phosphoenolpyruvate (DGº′ = -62 $kJmol^{-1}$) transfer of phosphoryl group to ADP to form pyruvate; reaction is metabolically irreversible (Q is far from K_{eq})

  - e.g. phosphagens, such as phosphocreatine and phosphoarginine
  - found in animal muscle cells

  - phosphocreatine acts as storage of phosphoryl group by the following reaction:

  creatine kinase

  phosphocreatine + ADP – – – – – –> creatine + ATP

  - phosphoarginine used in molluscs and arthropods

2) Nucleotidyl-group transfer

- e.g. synthesis of acetyl CoA - AMP is transferred to nucleophilic carboxylate group of acetate → acetyl group is transferred to sulfur atom of CoA

acetyl CoA
synthetase

ATP + acetate + CoA ------> AMP + acetyl CoA

3) **Thioesters**

- usually make ATP equivalents

  succinyl CoA + GDP + P_i --> succinate + GTP + HS-CoA

Reduced Coenzymes

Another class of energy-rich molecules.

Energy can be donated in oxidation-reduction reactions

$$A_{red} + B_{ox} \rightarrow A_{ox} + B_{red}$$

Electrons are transferred to oxidizing agents NAD^+ or FAD.

$NADH \rightarrow NAD^+$	ATP production in mitochondria
$NADP^+ \rightarrow NADPH$	Pentose phosphate pathway
$FMNH_2 \rightarrow FMN$	ETS electron carrier
$FADH_2 \rightarrow FAD$	ETS coenzyme

Cells Get Energy by Oxidizing Stored Fat and Sugar to Make ATP

Sugars, fats, and to some extent proteins have bond energy that can be released by oxidation

Some of this energy can be trapped in the bonds of ATP so that it can be used to supply energy for cellular processes

Sugars (i.e., glycogen) and proteins have about 4 kcalories per gram of stored energy, while fats (triglycerides) have about 9 kcal/gram

If glucose is heated in the presence of oxygen it will burn in a single step reaction, releasing 686 kcal/mole of free energy as heat

This amount of energy, released at one time, is too large for most biological reactions

Cells Oxidize Sugars in Small Steps in Water at Low Temperature

Cells "burn" glucose at low temperature

Glucose is oxidized to pyruvate in 10 steps, releasing small amounts of energy at each step

Pyruvate is further oxidized in most eukaryotic cells to CO_2 and water; this takes another 9 steps

Further steps in the electron transport system trap energy in ATP bonds

The final products of biological oxidation (CO2 and water) are the same as those obtained by high temperature burning, but much of the energy is trapped in ATP bonds

If you think of glucose as a tree, then biological oxidation is a bit like chopping the tree into logs of convenient length for future use

In Metabolism Carbon is Oxidized by Removing Hydrogen Atoms or Adding Oxygen Atoms

Loss of an electron, producing an ion, is called oxidation

If an electron is shifted away from an atom it is also called oxidation even if the electron is not completely removed

Oxygen is more electronegative than carbon (it has a greater affinity for electrons than carbon)

If carbon binds to oxygen the carbon is oxidized because electrons shift toward oxygen

In this reaction the oxygen is reduced

Hydrogen has less attraction for electrons than carbon

If carbon binds to a hydrogen the carbon is reduced and the hydrogen is oxidized

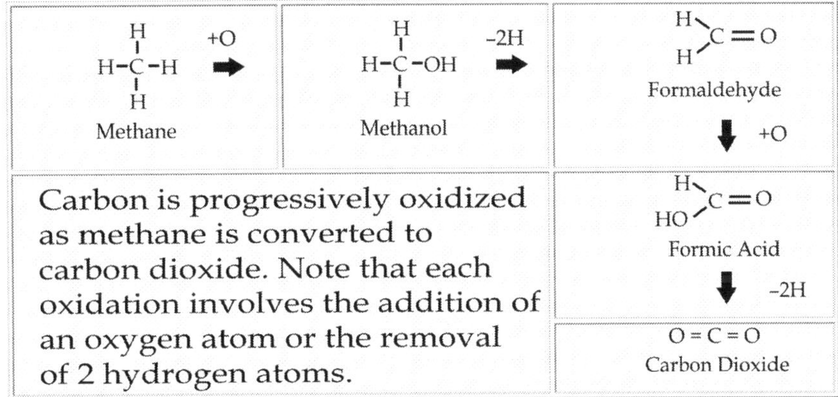

In Biological Oxidations Hydrogen Atoms are Usually Transferred to Coenzymes

In biochemistry H atoms stripped from carbons are transferred to coenzymes derived from B vitamins

The enzymes removing the Hs are called dehydrogenases

Coenzymes are present in small amounts and must be recycled to keep reactions going

NADH and FADH2 are produced in glycolysis and respiration

They are usually recycled (to NAD+ and FAD) by the Electron Transport System of the mitochondria

These reactions produce large amounts of ATP

Coenzyme	Source	Typical Reaction
Nicotinamide Adenine Dinucleotide	Niacin	H-C-O-H + NAD^+ ⬇ C=O + NADH + H^+
Flavin Adenine Dinucleotide	Riboflavin	H-C-H / H-C-H + FAD ⬇ C-H ‖ H-C + $FADH_2$
Nicotinamide Adenine Dinucleotide Phosphate	Niacin	$2H_2O + 2\ NADP^+$ ⬇ $2NADPH + 2H^+ + O_2$

An Overview of Energy Metabolism

The figure shows the central pathways supplying energy in plant, animal and many other types of cells

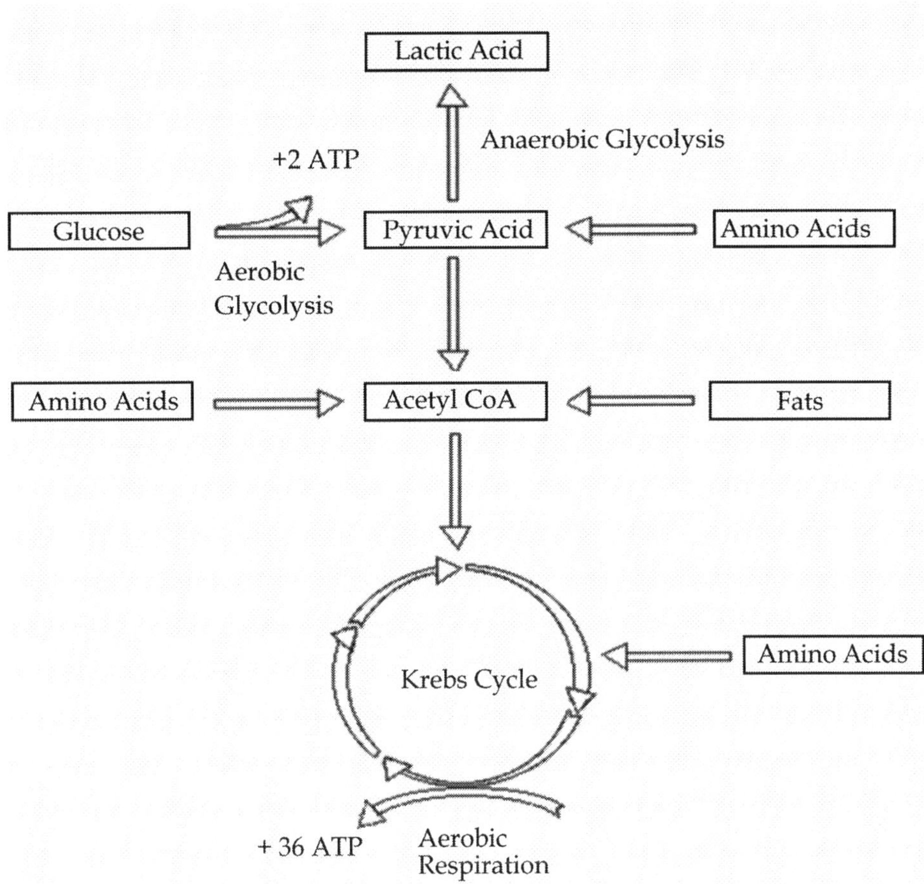

A. CARBOHYDRATE METABOLISM

- **Glycolysis**: splits glucose to pyruvate, which can be converted to lactate under anaerobi condition.
- **Gluconeogenesis**: converts pyruvate to glucose.
- **Glycogenesis**: synthesis of glycogen, carbohydrate fuel storage form.
- **Glycogenolysis**: breakdown of glycogen.
- **Pentose Phosphate Pathway (PPP) or HMP pathway**: produces NADPH for cell biosynthesis.
- **Kreb's TCA cycle or Citric Acid Cycle**: converts Acetyl CoA to CO_2 and ENERGY

Overviews of Glucose Metabolism

Glycolysis is going to be our first pathway, and it is arguably the most important and universal of the metabolic pathways. But before begin glycolysis let's take a brief look at how glucose (and carbohydrate generally) gets to the tissue from food intake.

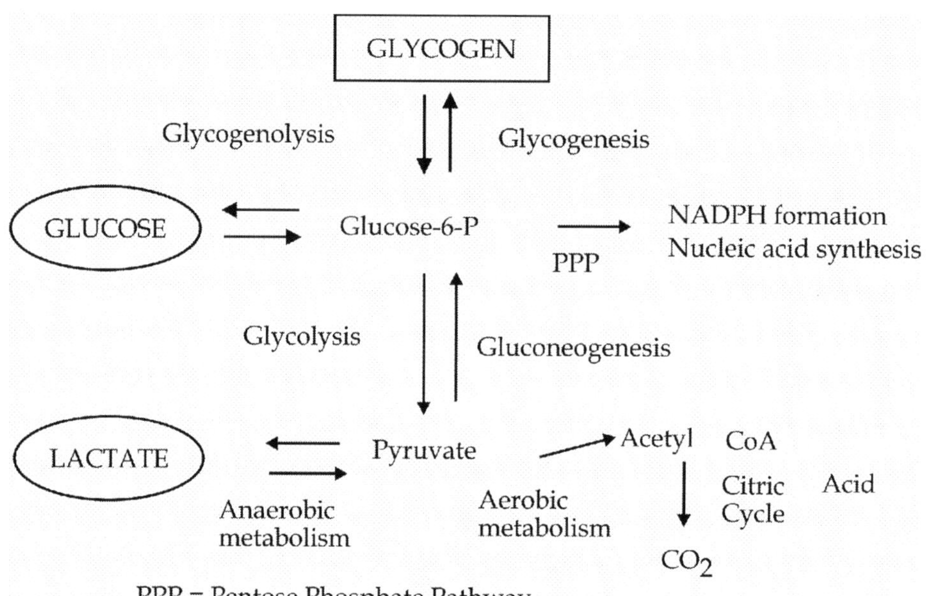

Digestion

- Initial, limited, breakdown of starch in mouth by salivary amylase
- Low pH of stomach inactivates salivary amylase, denatures many proteins etc.
- In duodenum digestive enzymes are added from the pancreas (pancreatic amylase, lipases, proteases), pH is neutralized, and detergents (bile acids) are added from gall bladder.
- Starches, Proteins, Lipids are all hydrolyzed in duodenum and small intestine where monomers are absorbed across intestinal epithelium. Amino acids and sugars are released into the portal vein to travel to the liver and then the rest of the body.
- First pass through the liver removes sugars and most of the amino acids for processing, storage with the release of glucose and fatty acids over time for use in the peripheral tissues.

Significance of glycolysis

Glycolysis is an almost universal central pathway of glucose catabolism occurring in the cytoplasm of all the tissues of biological systems leading to generation of energy in the form of ATP for vital activities.

It is the pathway through which the largest flux of carbon occurs in most cells.

In plants, glycolysis is the key metabolic component of the respiratory process, which generates energy in the form of ATP in cells where photosynthesis is not taking place.

Many types of anaerobic microorganisms are entirely dependent on glycolysis. Mammalian tissues such as renal medulla and brain solely dependent on glycolysis for major sources of metabolic energy.

Energy Yield

Two ATPs are used in glycolysis and four ATPs are synthesized for each molecules of glucose so that the net yield is two

ATPs per glucose. Under aerobic conditions, the two NADH molecules arising from glycolysis also yield energy via oxidative phosphorylation.

GLYCOLYSIS

Purpose: catabolism of glucose to provide ATPs and NADH molecules

Also provides building blocks for anabolic pathways.

Sequence of 10 enzyme-catalyzed reactions:

glucose → pyruvate 2 ATPs and 2 NADH produced

All enzymes (and reactions) are cytosolic.

Net reaction:

$$\text{glucose} + 2ADP + 2NAD^{+} + 2P_i \rightarrow 2\ \text{pyruvate} + 2ATP + 2NADH + 2H^{+} + 2H_2O$$

Can catabolize sugars other than glucose:

e.g. fructose → 2 glyceraldehyde 3-phosphate

e.g. lactose → glucose + galactose

galactose → glucose 1-phosphate → glucose 6-phosphate

e.g. mannose → mannose 6-phosphate → fructose 6-phosphate

Ten Steps of Glycolysis

1) glucose ® glucose 6-phosphate by hexokinase G = -8.0 kcal/mole Hexokinase also works on mannose and fructose at increased [ ]. Serves to trap glucose in the cell ® a phosphorylated molecule cannot leave

2) glucose 6-phosphate ® fructose 6-phosphate by glucose 6-phosphate isomerase

   Example of aldose ® ketose isomerization.

   Enzyme is very stereospecific.

   Reaction is near equilibrium in cell ® not a control point in glycolysis

3) fructose 6-phosphate ® fructose 1,6-bisphosphate by phosphofructokinase-1 (PFK-1)

   Reaction has G = -5.3 kcal/mole and is metabolically irreversible.

   Represents the first committed step in glycolysis.

4) fructose 1,6-bisphosphate ® dihydroxyacetone phosphate + glyceraldehyde 3-phosphate by fructose 1,6 bisphosphate aldolase.

5) DHAP ® glyceraldehyde 3-phosphate by triose phosphate isomerase

   Also catalyzes aldose ® ketose conversion.

   Rate is diffusion controlled (substrate is converted to product as fast as substrate is encountered).

6) glyceraldehyde 3-phosphate ® 1,3-bisphosphoglycerate by glyceraldehyde 3-phosphate dehydrogenase

   One molecule of NAD+ is reduced to NADH ® respiratory chain

7) 1,3 bisphosphoglycerate ® 3-phosphoglycerate

   Phosphoryl group transfer to ADP to form ATP.

   Because phosphate group comes from a substrate molecule, called **substrate level phosphorylation**

   First ATP-generating step of glycolysis.

8) 3-phosphoglycerate ® 2-phosphoglycerate by phosphoglycerate mutase

   Mutases are enzymes that transfer phosphoryl groups from one part of a substrate molecule to another.

9) 2-phosphoglycerate ® phosphoenolpyruvate (PEP) by enolase (forms double bond)

10) PEP ® pyruvate

Second time for substrate level phosphorylation.

Reaction is metabolically irreversible.

GLUCONEOGENESIS

- Synthesis of glucose from noncarbohydrate sources.
- Major precursors are lactate and alanine in the liver and kidney.
- lactate - active skeletal muscles
- glycerol - lipid catabolism
- amino acids - diet and protein catabolism
- Used to maintain blood glucose levels when glycogen supplies are low or depleted.
- Major site of occurrence is the liver, but also occurs in kidney.
- Designed to make sure blood glucose levels are high enough to meet the demands of brain and muscle (cannot do gluconeogenesis).
- NOT the reverse of glycolysis. Why?
- PFK, PK, and hexokinase catalyze metabolically irreversible steps.
- Solution: by-pass these steps, but use all the other enzymes.

1) Pyruvate → phosphoenolpyruvate

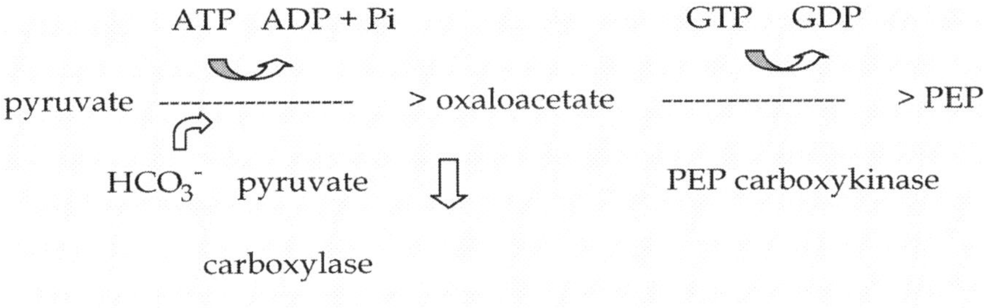

TCA Cycle

2) Fructose 1,6 bisphosphate $\xrightarrow{Pi}$ fructose 6-phosphate

fructose 1,5-bisphosphatase

glucose 6-phosphatase

3) Glucose 6-phosphate ——→ glucose

This enzyme is bound to ER membrane, but faces ER lumen.

GLUT7 transporter must transport glucose 6-phosphate into ER lumen.

Enzyme not found in membrane of brain or muscle ER.

Consequences?

Precursors for Gluconeogenesis

1) Lactate

Cori cycle - no net gain or loss of glucose

Anaerobic respiration of pyruvate.

2) Amino acids

3) Glycerol

glycerol kinase

glycerol ——→ glycerol 3-phosphate → DHAP

If glycerol 3-phosphate dehydrogenase is embedded in inner mitochondrial membrane, e- passed to ubiquinone.

If enzyme is cytosolic, NADH is also a product.

Regulation of Gluconeogensis

- Glycolysis and gluconeogenesis are reciprocally regulated.
- If both pathways were activated, e.g.

  Fructose 6-phosphate + ATP → Fructose 1,6-bisphosphate + ADP

  Fructose 1,6-bisphosphate + H_2O → Fructose 6-phosphate + P_i

  net reaction: ATP + H_2O → ADP + P_i
- Called **substrate cycle** → "burn" 4 ATPs for every 2 ATPs made (can be used to generate heat).

- Reason why enzymes are regulated → prevents this from happening.
- Two regulatory points are the two steps which had different enzymes.

  Fructose 1,6-bisphosphatase inhibited by AMP and Fructose 2,6-bisphosphate

  pyruvate carboxylase activated by acetyl CoA

Fate of Pyruvate After Glycolysis

Under anaerobic conditions, cells must be able to regenerate NAD^+ or glycolysis will stop.

Usually regenerated by oxidative phosphorylation, but that requires O_2.

There are 2 anaerobic pathways that use NADH and regenerate NAD+.

1) Alcoholic fermentation

   Conversion of pyruvate to ethanol

$$\text{pyruvate} \xrightarrow[\text{pyruvate decarboxylase}]{H^+ \quad CO_2} \text{acetaldehyde} \xrightarrow[\text{alcohol dehydrogenase}]{NADH \quad NAD^+} \text{ethanol}$$

glucose $+2P_i + 2ADP + 2H^+ \rightarrow 2$ ethanol $+ 2CO_2 + 2ATP + 2H_2O$

2) Lactate fermentation

$$\text{pyruvate} \xrightarrow[\text{lactate dehydrogenase}]{NADH + H^+ \quad NAD^+} \text{lactate}$$

   glucose $+2P_i + 2ADP \rightarrow 2$ lactate $+ 2ATP + 2H_20$

   Lactate causes muscles to ache.

   Also produced by bacterial fermentation of lactose.

3) Entry into citric acid cycle

   Glucose comes from 2 locations

Glycolysis is Regulated at Several Sites

The cell cannot store much ATP, so its production must be controlled by current needs

One way in which this is done is by negative feedback by products of the reactions

Example:Phosphofructokinase, which catalyzes 3rd reaction of glycolysis

Inhibited by high ATP levels and by citric acid (a Krebs cycle intermediate)

Activated by high levels of ADP

Hexokinase and pyruvate kinase, which catalyze the 1st and last steps of glycolysis, are also regulated enzymes

The Citric Acid Cycle

The **Tricarboxylic acid** cycle is in many ways the central pathway of metabolism, both catabolically and anabolically: it is involved in the breakdown and synthesis of a variety of compounds. The oxidative breakdown of the acetyl group of acetyl CoA.

Interconversion of metabolic intermediates: The TCA cycle has a central place in metabolism (even in anaerobic organisms) via its use to interconvert metabolites.

For the entire cycle, the production of 10 ATP/acetyl-CoA or 20 ATP/ Glucose.

Yields reduced coenzymes (NADH and QH_2) and some ATP (2).

Preparative step is oxidative decarboxylation involving coenzyme A.

Occurs in eucaryotic mitochondrion and prokaryotic cytosol.

How does the pyruvate get into the mitochondrion from the cytosol?

Pyruvate passes through channel proteins called **porins** (can transport molecules < 10,000 daltons) located in outer mitochondrial membrane.

To get from intermembrane space to matrix involves **pyruvate translocase** (symporter that also moves H^+ into matrix).

CONVERSION OF PYRUVATE TO ACETYL COA

Enzyme is pyruvate dehydrogenase complex, composed of three enzymes:

1) pyruvate dehydrogenase
2) dihydrolipoamide acetyltransferase
3) dihydrolipoamide dehydrogenase

Reaction occurs in 5 steps:

1) E_1 uses TPP as a prosthetic group and decarboxylates pyruvate → forms HETPP intermediate

2) E_1 then transfers acetyl group to oxidized lipoamide → acetyllipoamide

3) E_2 transfers acetyl group to coenzyme A to form acetyl CoA; dihydrolipoamide becomes reduced

4) E_3 reoxidizes lipoamide portion of E_2; prosthetic group of E_3 (FAD) oxidizes reduced lipoamide → $FADH_2$

5) NAD^+ is reduced by E_3-FADH → E_3-FAD + NADH + H^+

E2 acts like a crane by swinging substrate between protein complexes in enzyme.

Regulation of PDH complex:

Regulated by covalent modification by phosphorylation.

inactive = phosphorylated; active = dephosphorylated

E_1 inhibited at high [ATP]; inhibited at high [GTP]

activated by high [AMP], high [Ca^{2+}], high [pyruvate]

E_2 inhibited by high [acetyl CoA]

activated by high [CoA-SH]

E_3 inhibited by high [NADH]

activated by high [NAD^+]

The Citric Acid Cycle

Summary:

Composed of 8 reactions

4 carbon intermediates are regenerated

2 molecules of CO_2 released (6C→ 4C)

Most of energy stored as NADH and QH_2

1) Citrate synthase

Irreversible reaction

Acetyl CoA reacts with oxaloacetate → citrate and CoA

2) Aconitase

Citrate → isocitrate

3) Isocitrate dehydrogenase

Irreversible reaction

Substrate first oxidized ($2e^-$ and H^+ given to NAD^+), then decarboxylated

Isocitrate → ∞-ketoglutarate + CO_2 + NADH + H^+

4) ∞-ketoglutarate dehydrogenase complex

-ketoglutarate first decarboxylated, oxidized ($2e^-$ and H^+ given to NAD^+), and HS-CoA added

Product is succinyl CoA

Enzyme complex similar the PDH, but has dihydrolipoamide succinyltransferase instead of acetyltransferase.

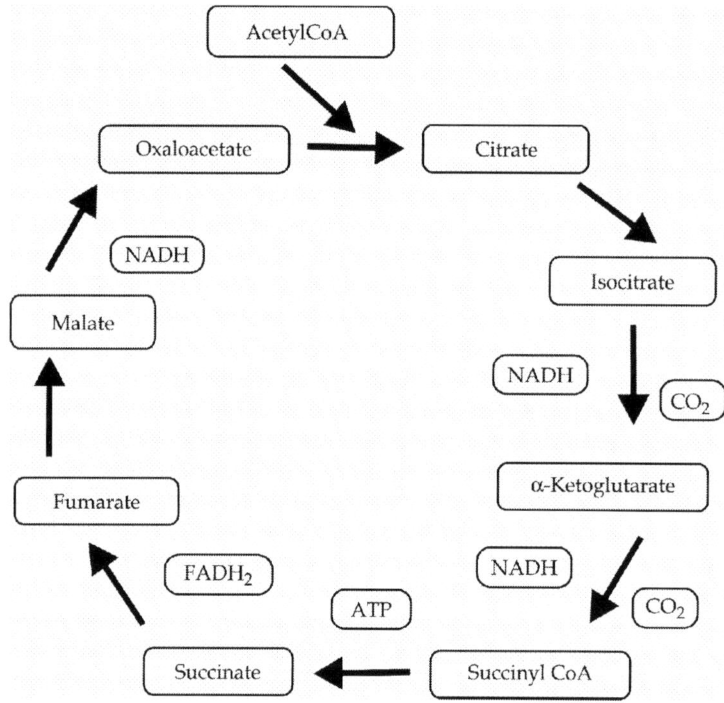

5) succinyl CoA synthetase or succinate thiokinase

succinyl CoA → succinate

Substrate has high energy thioester bond; that energy is stored as nucleoside triphosphate via substrate level phosphorylation

GDP +P_i → GTP mammals

ADP +P_i → ATP plants and bacteria

6) succinate dehydrogenase complex

Enzyme is embedded in inner mitochondrial membrane.

Has FAD covalently bound to it (prosthetic group).

Converts succinate → fumarate with generation of $FADH_2$ –> ETS

FAD is regenerated by reduction of a mobile molecule called **ubiquinone** (coenzyme Q) → QH_2.

7) fumarase

   fumarate → malate

8) malate dehydrogenase

   L-malate → oxaloacetate

   $2e^-$ and H^+ given to NAD^+ → NADH

Net reaction for citric acid cycle:

acetyl CoA + 3NAD+ + Q + GDP(ADP)+ P_i +$2H_2O$ → HS-CoA + 3NADH + QH_2 + GTP(ATP) + $2CO_2$ + $2H^+$

Regulation of the TCA Cycle

In order to understand the regulation of the TCA cycle, it is necessary to look at the ΔG values for the various reactions and the kinetic properties of the enzymes. Values for the non-equilibrium reactions are tabulated below:

There are 2 enzymes that are regulated:

1) Isocitrate dehydrogenase

   Allosterically activated by high [Ca^{2+}] and high [ADP]

   Allosterically inhibited by high [NADH]

2) α-Ketoglutarate dehydrogenase

   Allosterically activated by high [Ca^{2+}]

   Allosterically inhibited by high [NADH] and high [succinyl CoA]

ENTRY AND EXIT OF METABOLITES

Citrate, ∞-ketoglutarate, succinyl CoA, oxaloacetate lead to biosynthetic pathways.

Citrate → fatty acids and sterols in liver and adipocytes

(cleaved into acetyl CoA if needed)

∞-ketoglutarate → glutamate → amino acid synthesis or nucleotide synthesis

succinyl CoA → propionyl CoA → fatty acid synthesis

→ porphyrin synthesis

oxaloacetate → gluconeogenesis

→ asparate → urea synthesis, amino acids synthesis,

pyrimidine synthesis

Pathway intermediates must be replenished by **anapleurotic** reactions.

MITOCHONDRIAL ELECTRON TRANSPORT SYSTEM

The Electron Transport System

The inner mitochondrial membrane is protein rich. If carefully broken down, it is very rich in five protein complexes: I -IV are large protein complexes involved in electron transport, while V is the ATP sythatase driven by proton gradients

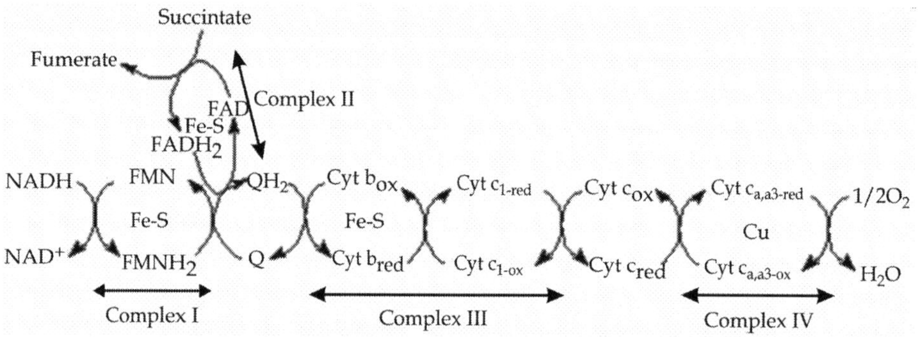

- Oxidative phosphorylation - process in which NADH and QH_2 are oxidized and ATP is produced.
- Enzymes are found in inner mitochondrial membrane in eukaryotes.
- In prokaryotes, enzymes are found in cell membrane.
- Process consists of 2 separate, but coupled processes:
  1) Respiratory electron-transport chain
     - Responsible for NADH and QH_2 oxidation
     - Final e- acceptor is molecular oxygen
     - Energy generated from electron transfer is used to pump H+ into intermembrane space from matrix → matrix becomes more alkaline and negatively charged.
  2) ATP synthesis
     - Proton concentration gradients represents stored energy
     - When H^+ are moved back across inner mitochondrial membrane through ATP synthase → ADP is phosphorylated to form ATP

Chemiosmotic Theory of ATP Production

Proposed by Peter Mitchell in 1961 (won Nobel Prize for this work).

Tenet: Proton concentration gradient serves as energy reservoir for ATP synthesis.

Proton concentration gradient also known as **proton motive force** (PMF).

Components of Electron Transport System

There are 5 protein complexes:

I) NADH-ubiquinone oxidoreductase

II) succinate-ubiquinone oxidoreductase

III) ubiquinol-cytochrome c oxidoreductase

IV) cytochrome c oxidase

V) ATP synthase

- Electrons flow through ETS in direction of increasing reduction potential.
- Two mobile electron carriers also involved: **ubiquinone** (Q) between complexes I or II and III, and **cytochrome c** between complexes III and IV.
- Electrons enter ETS 2 at a time from either NADH or succinate.

I - NADH-ubiquinone oxido reductase

Transfers $2e^-$ from NADH to Q as hydride ion (H-)

First electron transferred to FMN → $FMNH_2$ → Fe-S cluster → Q

Also pumps $4H^+/2e^-$ into intermembrane space

II - succinate-ubiquinone oxidoreductase

Transfers e^- from succinate to Q

First transferred to FAD → $FADH_2$ → 3 Fe-S clusters —> Q

Not enough energy to contribute to proton gradient via proton pumping

III - ubiquinol-cytochrome c oxidoreductase

Transfers e^- from QH_2 to cytochrome c facing intermembrane space

Composed of 9-10 subunits including 2 Fe-S clusters, cytochrome b_{560}, cytochrome b_{566}, and cytochrome c_1.

Transports $2H^+$ from matrix into intermembrane space

IV - Cytochrome c oxidase

Contains cytochromes a and a_3

Contributes to proton gradient in two ways:

1) pumps $2H^+$ for each pair of e^- transferred (per O_2 reduced)
2) consumes $2H^+$ when oxygen is reduced to $H_2O \rightarrow$ lowers $[H^+]_{matrix}$

   Carbon monoxide (CO) and cyanide (HCN) bind here

V - ATP synthase

Does not contribute to H^+ gradient, but helps relieve it

Also called F_OF_1 ATP synthase

F_1 component contains catalytic subunits

F_O component is proton channel that is transmembrane

Per ATP synthesized, $3H^+$ move through ATP synthase

oligomycin - antibiotic that binds to channel and prevents proton entry $\rightarrow$ no ATP synthesized

TRANSPORT OF MOLECULES ACROSS MITOCHONDRIAL MEMBRANE

- Inner mitochondrial membrane is impermeable to NADH and NAD^+.
- Must use a shuttle to regenerate NAD^+ for glycolysis; solution is to shuttle electrons across membrane, rather than NADH itself.
- There are two shuttles in operation:

  1) Glycerol phosphate shuttle

  - Found in insect flight muscles and mammalian cells in which high rates of oxidative phosphorylation must occur
  - Cytosolic glycerol 3-phosphate dehydrogenase converts DHAP to glycerol 3-phosphate
  - Converted back to DHAP by membrane-bound glycerol 3-phosphate dehydrogenase
  - Result is transfer to $2e^-$ to FAD $\rightarrow$ Q $\rightarrow$ complex III
  - Produces fewer ATP molecules (1.5 vs. 2) because complex I is bypassed

2) **Malate-Aspartate shuttle**
   - Found in liver and heart
   - Cytosolic NADH reduces oxaloacetate → malate → transported via dicarboxylate translocase into matrix
   - In matrix, malate → oxaloacetate → aspartate → transported out via glutamate-aspartate translocase
   - Converted back to oxaloacetate.......
   - No reduction in ATP yield

Must also be able to transport other metabolites into and out of matrix:

1) **ADP/ATP carrier or ADP/ATP translocase**

   Adenine nucleotide translocase which exchanges ADP and ATP (antiporter)

2) **P_i/H^+ carrier**

   Couples inward movement of P_i with symport of H^+ from gradient

REGULATION OF OXIDATIVE PHOSPHORYLATION

- Depends upon substrate availability and energy demands in the cell.
- Important substrates are NADH, O_2, and ADP.
- As ATP is used, more ADP is available, translocated through adenine nucleotide translocase → electron transport increases.
- Known as **respiratory control.**
- Helps to replenish ATP pool in the cell, which is kept nearly constant.
- Rates of glycolysis, citric acid cycle, and electron transport system are matched to a cell's ATP requirements.
- Proton gradient can be short-circuited to generate heat
- Found in brown adipose tissue in newborn mammals and animals that hibernate, and animals adapted to cold conditions
- A protein called **thermogenin** forms a proton channel in inner mitochondrial membrane → dissipates proton gradient, but electrons still flow → heat production
- Pathway is activated by fatty acids from triacylglycerol catabolism from epinephrine stimulation

Superoxide Production

Even though cytochrome oxidase and other proteins that reduce oxygen have been designed not to release O_2.- (superoxide anion), it still does happen.

Protonation of superoxide anion yields hydroperoxyl radical (HO_2.), which can react with another molecule to produce H_2O_2.

Enzyme superoxide dismutase catalyzes this reaction

$$O_2.- + O_2.- \xrightarrow[\text{superoxide dismutase}]{2H^+} H_2O_2 + O_2$$

Recent findings have indicated that superoxide dismutase mutations can cause **amyotrophic lateral sclerosis** (Lou Gehrig's disease), in which motor neurons in brain and spinal cord degenerate.

The hydrogen peroxide formed is scavenged by **catalase**:

$$H_2O_2 + H_2O_2 \xrightarrow[\text{catalase}]{} 2H_2O + O_2$$

Peroxidases catalyze an analogous reaction:

$$ROOH + AH_2 \xrightarrow[\text{Peroxidase}]{} ROH + H_2O + A$$

Energy Budget so far from 1 molecule of glucose is oxidized:

An accounting summary:

Reaction	Energy Product	Factor	ATP Equivalents (@ 2.5 ATP /NAD)	ATP Equivalents (@3 ATP/ NAD)
Glycolysis				
Hexokinase	ADP	1 x -1	- 1	-1
PFK	ADP	1 x-1	- 1	-1
GA-3-P DH	NADH	2 x 2.5 (1.5)*	5 (3)*	6 (4)*
PGA Kinase	ATP	2 x1	2	2
Pyruvate Kinase	ATP	2 x 1	2	2
Pyruvate DH Complex & Kreb's Cycle				
Pyruvate DH Complex	NADH	2 x 2.5	5	6
Isocitrate DH	NADH	2 x 2.5	5	6
2-oxoglutarate DH Complex	NADH	2 x 2.5	5	6
Succinyl-CoA Synthetase	GTP	2 x 1	2	2

Contd...

Succinate DH	$FADH_2$	2 x 1.5	3	4
Malate DH	NADH	2 x 2.5	5	6
TOTAL=	32 (30)*	38 (36)*		

* In some tissues (insect flight muscle, fast twitch muscle) the reducing equivalents of NADH must be pumped against a gradient at a cost of 1 ATP (it is used to make $FADH_2$).

SUMMARY OF THREE CYCLES (CONSIDERING 1NADH = 3 ATP)

	ATPs	NADHs	FADH2s	ATPs From ETC	Total ATPs
Glycolysis	2	2	0	6	8
Decarboxylation of Pyruvate	0	2	0	6	6
Krebs Cycle	2	6	2	22	24
Total	4	10	2	34	38

Pentose Phosphate Pathway (HMP Pathway)

- Provides NADPH (serves as e- donor) and forms ribose 5-phosphate (nucleotide synthesis).
- Pathway active is tissues that synthesize fatty acids or sterols because large amounts of NADPH needed.
- In muscle and brain, little PPP activity.
- All reactions are cytosolic.
- Divided into 2 stages:

The Pentose Phosphate Pathway is an alternate pathway for glucose oxidation which is used to provide reducing equivalents in support of biosynthesis.

Thus although it involves the catabolism of glucose, it is generally going to be active only when anabolism is taking place.

This pathway is usually treated in two parts:

the *oxidative* portion, and the *sugar interconversions* portion.

Regulation of pentose phosphate pathway

- Controlled by levels of $NADP^+$.
- Controlled step is dehydrogenation of glucose 6-phosphate to 6-phosphogluconolactone.
- Enzyme stimulated by high [$NADP^+$].
- Nonoxidative branch controlled primarily by substrate availability.

GLYOXYLATE CYCLE

Modification of citric acid cycle.

Anabolic pathway in plants, bacteria, yeast.

Takes 2 carbon compounds and converts them to glucose.

Common in plants which store energy reserves as oils, but must be converted to carbohydrates during germination.

In eucaryotes, a glyoxysome is a special organelle where this occurs.

Comparing the Energy Capture of Glycolysis and Respiration

Aerobic metabolism of glucose traps nearly 20 times more energy than anaerobic metabolism

Most of the energy captured from the Krebs cycle is in the bonds of the coenzymes, NADH and FADH2

In the electron transport chain the coenzyme energy is "cashed in" for ATP- a little like a currency exchange

The exchange rate is different for the different coenzymes

FADH2 gives 2 ATPs

NADH made in the mitochondria gets 3 ATPs

NADH made outside the mitochondria (in glycolysis) gets only 2 ATPs (some energy is lost in transporting this NADH into the mitochondria)

If the glucose is totally oxidized, using glycolysis, decarboxylation and the Krebs cycle, net 36 ATPs are generated per glucose, compared to only 2 if glycolysis alone is used

The great increase of energy generated by the mitochondria makes possible complex multicellular life forms

GLYCOGEN METABOLISM

- Glycogen stored in muscle and liver cells.
- Important in maintaining blood glucose levels.
- Glycogen structure: α 1,4 glycosidic linkages with α 1,6 branches.
- Branches give multiple free ends for quicker breakdown or for more places to add additional units.

Glycogen Degradation

Glycogenolysis

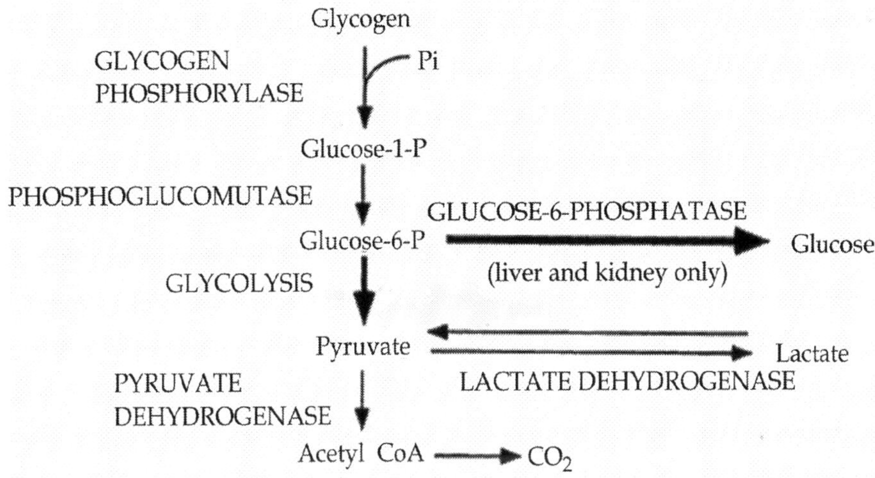

Glycogen is broken down using **Phosphorylase** to phosphorylize off glucose residues:

$$(\text{Glucose})_n + P_i \text{ ® } (\text{Glucose})_{n-1} + \text{G-1-P}$$

- Glucose residues of starch and glycogen released through enzymes called starch phosphorylases and glycogen phosphorylases.
- Catalyze **phosphorolosis**:

  polysaccharide $+P_i \rightarrow$ polysaccharide$_{(n-1)}$ + glucose 1-phosphate
- Pyridoxal phosphate (PLP) is prosthetic group in active site of enzyme; serves as a proton donor in active site.
- Allosterically inhibited by high [ATP] and high [glucose 6-phosphate].
- Allosterically activated by high [AMP].
- Sequentially removes glucose residues from nonreducing ends of glycogen, but stops 4 glucose residues from branch point —> leaves a **limit dextran**.
- Limit dextran further degraded by glycogen-debranching enzyme (glucanotransferase activity) which relocated the chain to a free hydroxyl end.
- Amylo-1,6-glucosidase activity of debranching enzyme removes remaining residues of chain.

- This leaves substrate for glycogen phosphorylase.
- Each glucose molecule released from glycogen by debranching enzyme will yield 3 ATPs in glycolysis.
- Each glucose molecule released by glycogen phosphorylase will yield 2 ATPs in glycolysis.
- Why?
  - ATP not needed in first step because glucose 1-phosphate already formed.

phosphoglucomutase

glucose 1-phosphate ⟶ glucose 6-phosphate

1) In liver, kidney, pancreas, small intestine,

glucose 6-phosphatase

glucose 6-phosphate ⟶ glucose + P_i

Glycogen Control

Since the Glycogen synthase and phosphorylase reactions are in opposition we need a control system. Glycogen storage/release strategies vary widely with different tissues. In liver, glycogen is used to provide glucose to the serum between meals - it serves a homeostatic function. Glycogen control in liver is thus designed to breakdown and release glycogen when serum [glucose] is low and synthesize glycogen when serum [glucose] is high.

Glycogen Synthesis

Glycogenesis

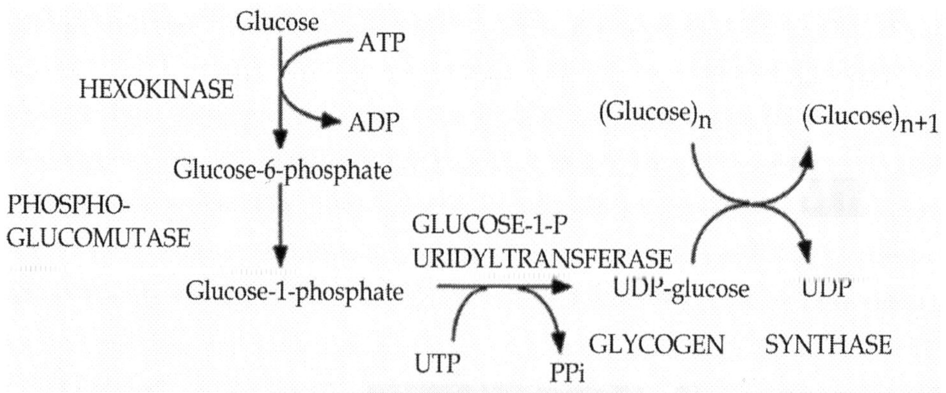

- Not reverse of glycogen degradation because different enzymes are used.
- About 2/3 of glucose ingested during a meal is converted to glycogen.
- First step is the first step of glycolysis:

hexokinase

glucose ⟶ glucose 6-phosphate

- There are three enzyme-catalyzed reactions:

phosphoglucomutase

glucose 6-phosphate ⟶ glucose 1-phosphate

glucose 1-phosphate ⟶ UDP-glucose (activated form of glucose)

glycogen synthase

UDP-glucose ⟶ glycogen

- Glycogen synthase cannot initiate glycogen synthesis; requires preexisting primer of glycogen consisting of 4-8 glucose residues with α (1,4) linkage.
- Protein called **glycogenin** serves as anchor; also adds 7-8 glucose residues.
- Addition of branches by **branching enzyme** (amylo-(1,4 –> 1,6)-transglycosylase).
- Takes terminal 6 glucose residues from nonreducing end and attaches it via a(1,6) linkage at least 4 glucose units away from nearest branch.

Regulation of Glycogen Metabolism

Mobilization and synthesis of glycogen under hormonal control.

Three hormones involved:

1) Insulin

- 51 amino acids protein made by cells of pancreas.
- Secreted when [glucose] high → increases rate of glucose transport into muscle and fat via GLUT4 glucose transporters.
- Stimulates glycogen synthesis in liver.

2) Glucagon

- 29 amino acids protein secreted by cells of pancreas.
- Operational under low [glucose].
- Restores blood sugar levels by stimulating glycogen degradation.

3) Epinephrine

- Stimulates glycogen mobilization to glucose 1-phosphate → glucose 6-phosphate.
- Increases rate of glycolysis in muscle and the amount of glucose in bloodstream.
- Occurs in response to fight-or-flight response.
- Binds to b-adrenergic receptors in liver and muscle and b_1 receptors in liver cells.
- Binding of epinephrine or glucagon to b receptors activates adenylate cyclase, which is a membrane-traversing enzyme that converts ATP → cAMP → activates protein kinase A.
- Binding of epinephrine to b_1 receptors activates IP_3 pathway → protein kinase C → phosphorylation of insulin receptors → insulin cannot bind.

Regulation of glycogen phosphorylase and glycogen synthase

- Reciprocal regulation.
  - Glycogen synthase -P → inactive form (b).
  - Glycogen phosphorylase-P → active (a).
- When blood glucose is low, protein kinase A activated through hormonal action of glucagon → glycogen synthase inactivated and phosphorylase kinase activated → activates glycogen phosphorylase → glycogen degradation occurs.
- Phosphorylase kinase also activated by increased [Ca^{2+}] during muscle contraction.
- To reverse the same pathway involves protein phosphatases, which remove phosphate groups from proteins → dephosphorylates phosphorylase kinase and glycogen phosphorylase (both inactivated), but dephosphorylation of glycogen synthase activates this enzyme.

- Protein phosphatase-1 activated by insulin → dephosphorylates glycogen synthase → glycogen synthesis occurs.
- In liver, glycogen phosphorylase a inhibits phosphatase-1 → no glycogen synthesis can occur.
- Glucose binding to protein phosphatase-1 activated protein phosphatase-1 → it dephosphorylates glycogen phosphorylase → inactivated → no glycogen degradation.
- Protein phosphatase-1 can also dephosphorylate glycogen synthase → active.

INDEED

- *Active* - dephosphorylated
- *Inactive* - phosphorylated
- Glycogen Synthase-i: *independent* (i) of glucose-6-phosphate for its activity.
- Glycogen Synthase-d: *dependence* (d) on *glucose-6-phosphate*, mechanism for storing glucose when overabundance is signalled by a build-up of glucose-6-phosphate.

B. LIPID METABOLISM

Pathways of the Fat System

Lipolysis: hydrolysis of triacylglycerol to glycerol and free fatty acids

β-Oxidation: mitochondrial oxidation of fatty acids ® ENERGY!

Ketogenesis: synthesis of ketone bodies

Lipogenesis: fatty acid synthesis

Esterification: attachment of fatty acids to glycerol to form triglycerides

Cholesterolgenesis: synthesis of cholesterol

Steroidogenesis: synthesis of steroid hormones from cholesterol

Interactions of Fat Metabolism Pathways

Fat metabolism in specific tissues

- **Liver**: oxidizes fats, produces ketones, exports triglycerides
- **Muscle**: uses fats and ketones as an energy source
- **Adipose Tissue**: stores fats as triacylglycerol and releases fatty acids as needed for energy

- **Mature RBC's** (no mitochondria) / **Brain** (lack of enzymes): do not use fats for energy

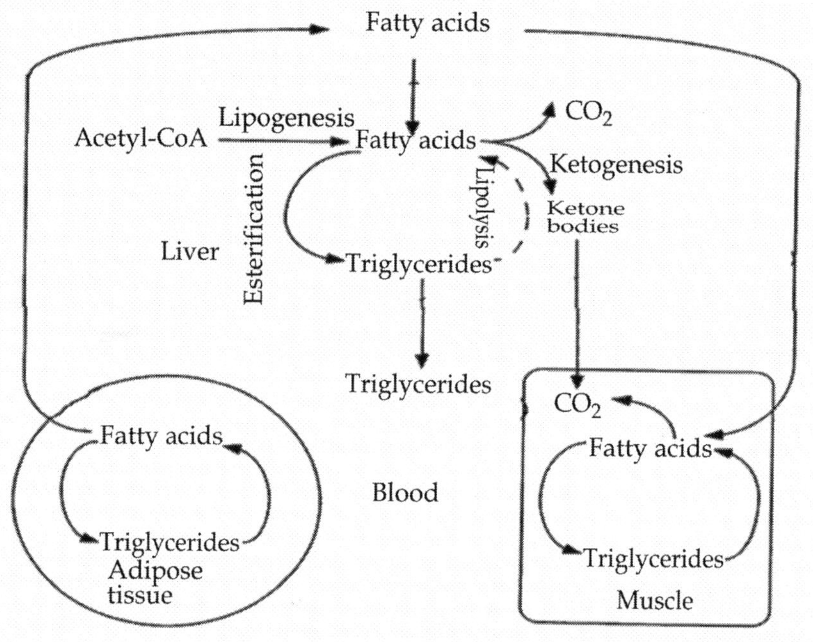

Fatty acids have four major physiologic roles in the cell

- Building blocks of phospholipids and glycolipids
- Added onto proteins to create lipoproteins, which targets them to membrane locations
- Fuel molecules - source of ATP
- Fatty acid derivatives serve as hormones and intracellular messengers

Absorption and Mobilization of Fatty Acids

- Most lipids are triacylglycerols, some are phospholipids and cholesterol.
- Digestion occurs primarily in the small intestine.
- Fat particles are coated with bile salts (amphipathic) from gall bladder.
- Degraded by **pancreatic lipase** (hydrolyzes C-1 and C-3 $\rightarrow$ 2 fatty acids and 2-monoacylglycerol).

- Can then be absorbed by intestinal epithelial cells; bile salts are recirculated after being absorbed by the intestinal epithelial cells.
- In the cells, fatty acids are converted by fatty acyl CoA molecules.
- Phospholipids are hydrolyzed by pancreatic phospholipases, primarily **phospholipase A2**.
- Cholesterol esters are hydrolyzed by **esterases** to form free cholesterol, which is solubilized by bile salts and absorbed by the cells.
- Lipids are transported throughout the body as **lipoproteins**.
- Lipoproteins consist of a lipid (tryacylglycerol, cholesterol, cholesterol ester) core with amphipathic molecules forming layer on outside.

Lipoproteins

- Both transported in form of lipoprotein particles, which solubilize hydrophobic lipids and contain cell-targeting signals.
- Lipoproteins classified according to their densities:
  - **chylomicrons** - contain dietary triacylglycerols
  - **chylomicron remnants** - contain dietary cholesterol esters
  - **very low density lipoproteins** (VLDLs) - transport endogenous triacylglycerols, which are hydrolyzed by lipoprotein lipase at capillary surface
  - **intermediate-density lipoproteins** (IDL) - contain endogenous cholesterol esters, which are taken up by liver cells via receptor-mediated endocytosis and converted to LDLs
  - **low-density lipoproteins** (LDL) - contain endogenous cholesterol esters, which are taken up by liver cells via receptor-mediated endocytosis; major carrier of cholesterol in blood; regulates *de novo* cholesterol synthesis at level of target cell
  - **high-density lipoproteins** - contain endogenous cholesterol esters released from dying cells and membranes undergoing turnover

Storage of Fatty Acids

- Triacylglycerols are transported as chylomicrons and VLDLs to adipose tissue; there, they are hydrolyzed to fatty acids, which enter adipocytes and are esterified for storage.
- Mobilization is controlled by hormones, particularly epinephrine, which binds to [illegible]-adrenergic receptors on adipocyte membrane →

protein kinase A activated → phosphorylates hormone-sensitive lipase → converts triacylglycerols to free fatty acids and monoacylglycerols.

- Insulin inhibits lipid mobilization (example of reciprocal regulation).
- Monoacylglycerols formed are phosphorylated and oxidized to DHAP (intermediate of glycolysis and gluconeogenesis).

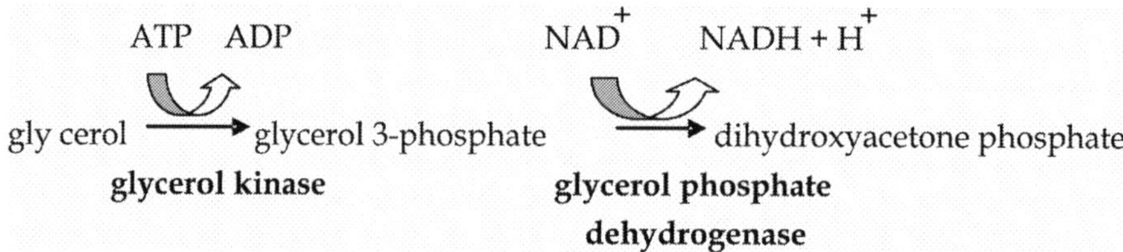

Can be converted to glucose (gluconeogenesis) or pyruvate (glycolysis) in the liver.

Oxidation of Fatty Acids

Fatty acids obtained by hydrolysis of fats undergo different oxidative pathways designated as (a), beta (b) and omega (w) pathways.

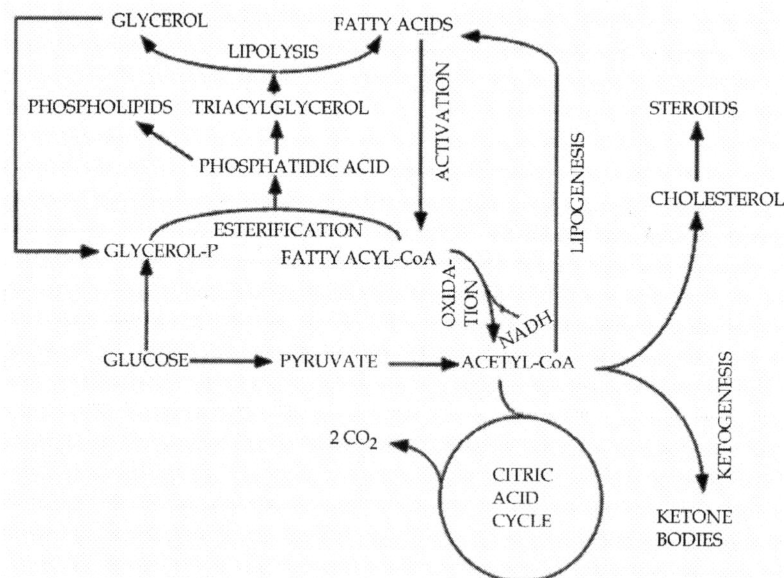

α-oxidation

α-Oxidation of fatty acids has been found in certain tissues especially in brain tissue of mammals and plant systems.

It does not require CoA intermediates and no high energy phosphates are generated. This type of oxidation results in the removal of one carbon at a time from the carboxyl end of the fatty acid.

The physiological role of a oxidation in plants is not yet fully established but it has been suggested that it may be involved in the degradation of long chain fatty acids as observed in many animal tissues.

α Oxidation is clearly the main source of the odd carbon fatty acids and their derivatives that occur in some plant lipids.

In this process, sequential removal of one carbon at a time from free fatty acids of chain length ranging from C13 to C18 occur.

ω-oxidation

ω Oxidation is normally a very minor pathway brought about by hydroxylase enzymes involving cytochrome P 450 in the endoplasmic reticulum.

Fatty acids with oxygen function (alcoholic or carboxyl) at the methyl terminal end (ω end) are formed by w oxidation and frequently occur as constituents of cutin and suberin.

The requirements for the oxygenase mediated conversion of a w methyl fatty acyl CoA into a w hydroxymethyl fatty acyl CoA are molecular oxygen, reduced pyridine nucleotide and a non-heme iron protein in higher plants.

β Oxidation of fatty acids

In 1904, Franz Knoop made a critical contribution to the elucidation of the mechanism of fatty acid oxidation and demonstrated that most of the fatty acids are degraded by oxidation at the b carbon.

β Oxidation of fatty acids takes place in mitochondria.

Fatty acids are activated before they enter into mitochondria for oxidation.

Free fatty acids are introduced into the cytosol, but β-oxidation occurs in the mitosol.

- Fatty acids are degraded by oxidation of the α-carbon by **β-oxidation.**
- Pathway that removes 2-C units at a time → acetyl CoA → citric acid cycle → ATP
- There are three stages in β-oxidation:
    - Activation of fatty acids in cytosol catalyzed by acyl CoA synthetase; two high energy bonds are broken to produce AMP
    - Transport of fatty acyl CoA into mitochondria via carnitine shuttle

- β-oxidation - cyclic pathway in which many of the same enzymes are used repeatedly

Two situations occur

- **Short to medium length fatty acids** are permeable to the mitochondrial membrane. They are activated to fatty acyl CoA derivatives in the mitochondrial matrix by **Butyryl-CoA Synthetase**:

b-Oxidation of Free Fatty Acids With an Even Number of Carbons

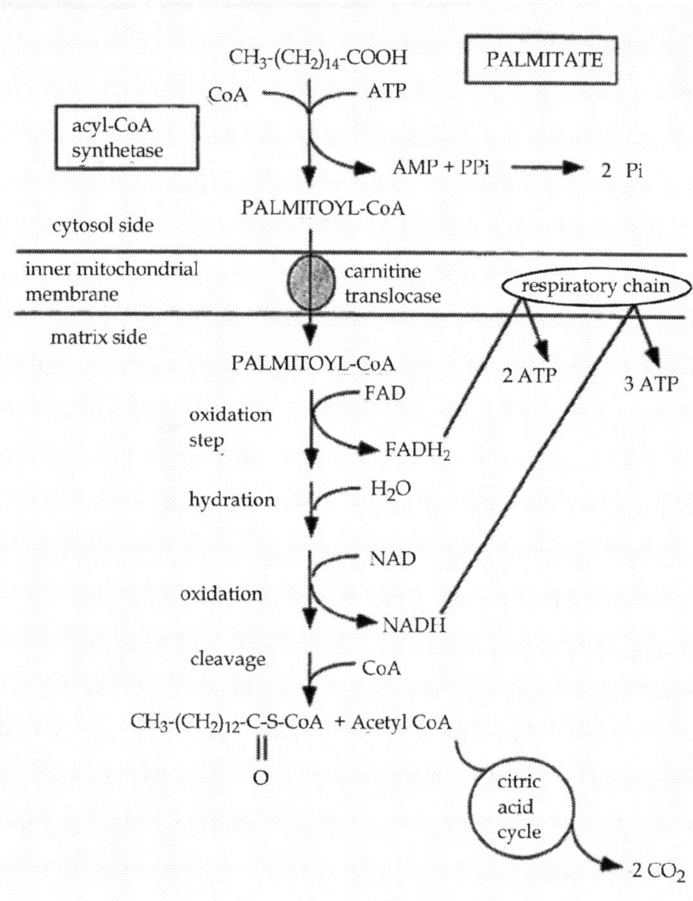

Energetics of β oxidation

The energetics or the energy conserved in terms of ATP by oxidation of a molecule of palmitic acid is given below:

Palmitic acid (16 carbons) undergoes β-oxidation forming eight molecules of acetyl CoA by undergoing seven β-oxidation spirals. When one cycle of β oxidation takes place, one molecule of FADH2, one molecule of NADH and one molecule of acetyl CoA are produced. Electrons from these reducing equivalents (FADH2 and NADH) are transported through the respiratory chain in mitochondria with simultaneous regeneration of high-energy phosphate bonds. Mitochondrial oxidation of FADH2 eventually results in the net formation of about 1.5 ATP. Likewise, oxidation of electrons from NADH yields 2.5 molecules of ATP. Hence, a total of four ATP molecules are formed per cycle and ten molecules of ATP are formed through Krebs's cycle from each molecule of acetyl CoA.

8 Acetyl CoA through TCA cycle yield (8x10)	= 80 ATP
7 β oxidation spiral reactions yield (7x4)	= 28 ATP
Total	108 ATP
ATP utilized in the initial step	= 2 ATP

Hence, complete oxidation of palmitic acid yields **106 ATP.**

β-Oxidation of unsaturated fatty acids

- First the double bond can occur in the wrong position, and
- Second, unsaturated fatty acids are commonly cis-, whereas *beta*-oxidation uses trans-fatty acids.

Two new enzymes are required to handle these situations: Enoyl-CoA isomerase (isomerizes a *cis*-3,4-double bond to a *trans*-2,3-double bond), and 2,4-Dienoyl-CoA reductase (reduces the *cis*-4,5-double bond in the *trans*-2,3-*cis*-4,5-dienoyl-CoA derivative formed during *beta*-oxidation). The resulting products are then broken down by the *beta*-oxidation enzymes.

Conditions Favoring Ketoacidosis

Relative or absolute deficiency of insulin

↓

Mobilization of free fatty acids from fat depots

↓

Increased delivery of free fatty acids to the liver

↓

Increased uptake and oxidation of free fatty acids by the liver

↓

Accelerated production of ketone bodies by the liver

Ketone Body Formation in Liver

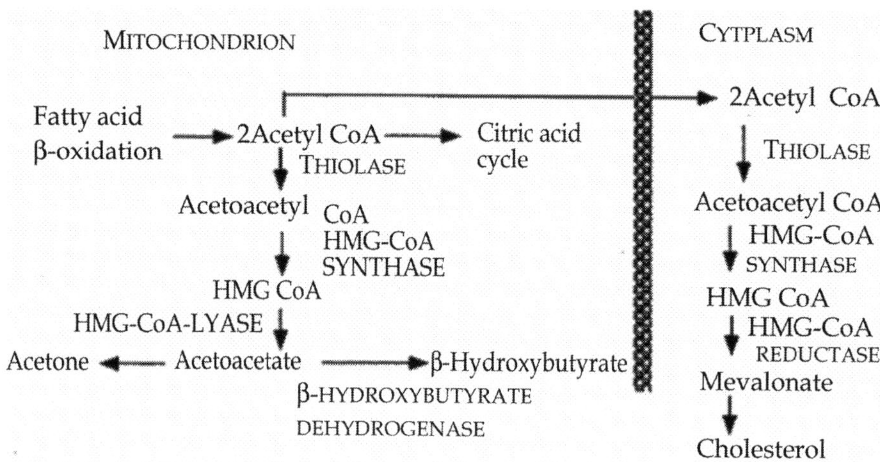

Normal prevention of ketoacidosis

Insulin whose release is promoted by ketone bodies, inhibits lipolysis to decrease the supply of fatty acids and thus curtail ketogenesis —> prevent ketoacidosis

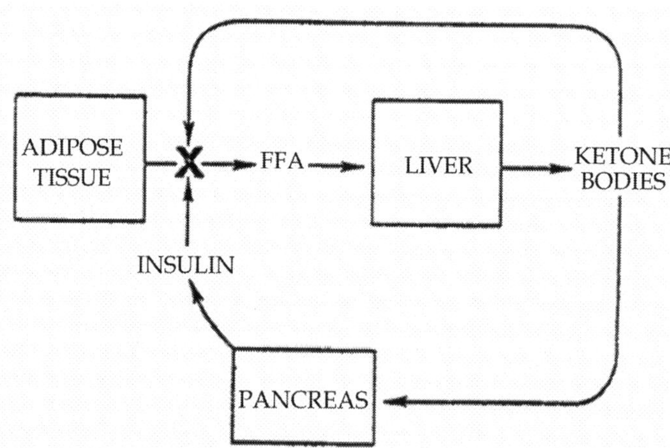

Ketone Body Oxidation: Long-term starvation or ketoacidosis

Tissues that can use ketones as "fuel": **brain, muscle, kidney, intestine**

β-hydroxybutyrate + $NAD^+ \rightarrow$ NADH + acetoacetate

acetoacetate + succinyl CoA $\rightarrow$ acetoacetyl CoA + succinate

acetoacetyl CoA + CoA $\rightarrow$ 2 acetyl CoA

Regulation of Fatty Acid Oxidation

- Already talked about fatty acid mobilization via epinephrine.
- Net result is high concentrations of acetyl CoA and NADH via α-oxidation.
- Both molecules allosterically inhibit pyruvate dehydrogenase complex.
- Most of acetyl CoA produced goes to Krebs cycle; during periods of fasting, excess acetyl CoA is produced, too much for Krebs cycle.
- Also in diabetes, oxaloacetate is used to form glucose by gluconeogenesis $\rightarrow$ concentration of oxaloacetate is lowered.
- Result is the diversion of acetyl CoA to form acetoacetate and 3-hydroxybutyrate; these two molecules plus acetone are known as **ketone bodies**.
- Acetoacetate is formed via the following reactions:

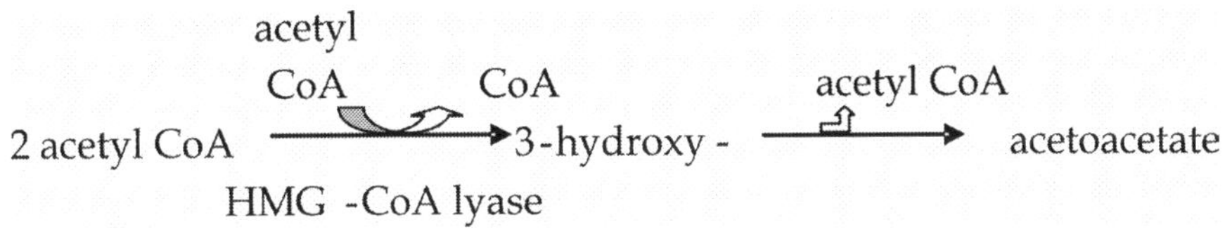

acetyl CoA (in), CoA (out), HMG -CoA lyase

2 acetyl CoA → 3-hydroxy - → (acetyl CoA) acetoacetate

- The major site of ketone body synthesis is the liver, within the mitochondrial matrix $\rightarrow$ transported to the bloodstream.
- Acetoacetate and 3-hydroxybutyrate are used in respiration and are important sources of energy.
- Cardiac muscle and the renal cortex perferentially use acetoacetate over glucose.
- Glucose is used by brain and RBCs; in brain, ketone bodies substitute for glucose as fuel because the brain cannot undergo gluconeogenesis.
- Acetoacetate can be converted to acetyl CoA and oxidized in citric acid cycle only in nonhepatic tissues.

Diabetes (insulin-dependent diabetes mellitus; IDDM)

Decreased insulin secretion by beta cells of pancreas; could be caused by viruses (?)

Juvenile onset

Patients are thin, hyperglycemic, dehydrated, polyuric (pee a lot), hungry, thirsty

In these patients, glycogen mobilization, gluconeogenesis, fatty acid oxidation occurs → massive ketone body production; also, some of the glucose is in urine (tends to pull water out of body) → **diabetic ketoacidosis**

Fatty Acid Synthesis

- Synthesis takes place in the cytosol, in contrast with degradation or oxidation, which occurs in the mitochondrial matrix.
- Intermediates in fatty acid synthesis are covalently linked to the sulfhydryl group of an acyl carrier protein (ACP) whereas intermediates in fatty acid breakdown are bonded to coenzyme A.
- The enzymes of fatty acid synthesis in animals are joined in a single polypeptide chain called fatty acid synthase. In contrast, the degradative enzymes do not seem to be associated. Plants employ separate enzymes to carry out the biosynthetic reactions.
- The reductant in fatty acid synthesis is NADPH, whereas the oxidants in fatty acid oxidation are NAD+ and FAD.

Lipogenesis (Fatty acid synthesis)

Pyruvate-malate cycle

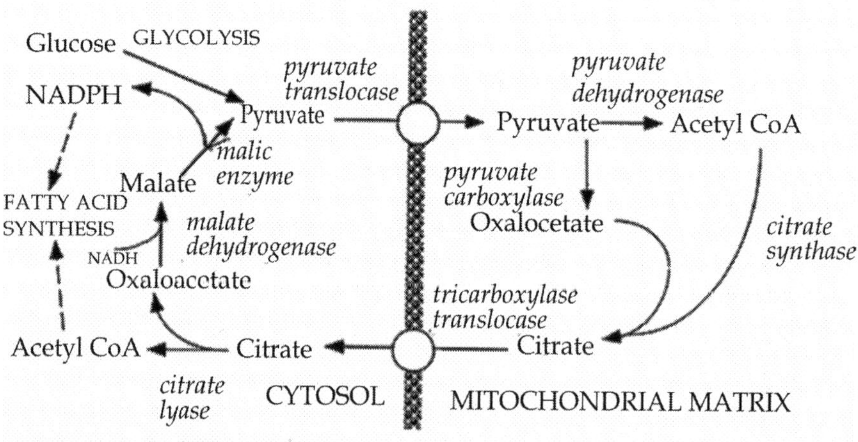

Now, let fatty acid biosynthesis starting from glucose. For that look at the integration of the various pathways involved: Glycolysis, hexose monophosphate shunt, Pyruvate-malate shuttle, and Fatty acid biosynthesis. It requires reducing equivalents, redox balance and provision of required cytosolic ATP's as well as carbon source.

The following steps are involved in fatty acid biosynthesis in cytosol

Two reactions enable the synthesis pathway:

- Acetyl-CoA is "activated" by the addition of a carbon dioxide, and
- NADPH is substituted as a more powerful reducing agent than $FADH_2$.

Summery of fatty acid biosynthesis

Acetyl CoA + 7 malonyl CoA + 14 NADPH + 20 H+ ⟶
Palmitate + 7 CO2 + 14 NADP+ + 8 CoASH + 6 H2O

The equation for the synthesis of the malonyl CoA used in the above reaction is

7 Acetyl CoA + 7 CO2 + 7 ATP ⟶ 7 malonyl CoA + 7ADP + 7 Pi + 14 H+

The overall reaction for the synthesis of palmitate is

8 Acetyl CoA + 7 ATP + 14 NADPH + 6H+ ⟶
Palmitate + 14 NADP + 8 CoASH + 6 H2O + 7 ADP + 7 Pi

Fatty acid synthesis and degradation are reciprocally regulated so that both are not simultaneously active.

Regulation of Lipogenesis

Enzyme		Regulatory agent	Effect
Acetyl CoA carboxylase	Short-term	Insulin	Stimulation
		Gluagon	Inhibition
	Long-term	High-carbohydrate, low-fat diet	↑enzyme synthesis
		High-fat diet	↓enzyme synthesis
		Fasting	↓enzyme synthesis
Fatty acid synthase		High-carbohydrate, low-fat diet	↑enzyme synthesis
		High-fat diet	↓enzyme synthesis
		Fasting	↓enzyme synthesis

Conditions Favoring Lipogenesis

increased glucokinase activity in liver - *increases metabolism of glucose as a fatty acid precursor*

decreased fatty acid availbility - *reduces inhibition of lipogenesis*

activation of acetyl CoA carboxylase - *increases production of malonyl CoA;*

increased flux through pentose shunt - *produces NADPH for fat synthesis*

hypercaloric high carbohydrate or high protein, low-fat diet - *provides precusors for fat synthesis*

Comparison of Fatty Acid β-Oxidation and Synthesis

Parameter	Oxidation	Synthesis
Intracellular location	Mitochondria	Cytoplasm
Coenzymes	FAD, NAD+	NADPH
Bicarbonate dependence	No	Yes
Citrate activation	No	Yes
Acyl CoA inhibition	No	Yes
Malonyl CoA inhibition	Yes	No
Highest activity	Fasting	Carbohydrate fed

Regulation of Fatty Acid Synthesis

- Metabolism of fatty acids is under hormonal regulation by glucagons, epinephrine, and insulin.
- Fatty acid synthesis is maximal when carbohydrate and energy are plentiful.
- Important points of control are release of fatty acids from adipocytes and regulation of carnitine acyltransferase I in the liver.
- High insulin levels also stimulate formation of malonyl CoA, which allosterically inhibits carnitine acyltransferase I → fatty acids remain in cytosol and are not transported to mitochondria for oxidation.
- Key regulatory enzyme is **acetyl-CoA carboxylase** (catalyzes first committed step in fatty acid synthesis).
- Insulin stimulates fatty acid synthesis and inhibits hydrolysis of stored triacylglycerols.
- Glucagon and epinephrine inhibit fatty acid synthesis (enzyme is phosphorylated by protein kinase A; removal of phosphate group catalyzed by protein phosphatase 2A).
- Citrate is an allosteric activator, but its biological relevance has not been established.
- Fatty acyl CoA acts as an inhibitor.
- Palmitoyl CoA and AMP are allosteric inhibitors.

Synthesis of Eicosanoids

- Precursors for eicosanoids are 20-carbon polyunsaturated fatty acids such as arachidonate.
- Part of inner leaflet of cell membrane.
- There are two classes of eicosanoids:

  1) prostaglandins and thromboxanes

     Synthesized by enzyme cyclooxygenase

     Localized molecules such as thromboxane A_2, prostaglandins, prostacyclin ae produced.

     Thromboxane A_2 leads to platelet aggregation and blood clots → reduced blood flow in tissues.

     Aspirin binds irreversibly to Cyclooxygenase enzymes and prevents prostaglandin synthesis.

  2) leukotrienes

     Produced by lipoxygenases.

     Products were once called "slow-acting substances of anaphylaxis", responsible for fatal effects of some immunizations.

Regulation of Fatty Acid Metabolism

Fatty acid metabolism is regulated both hormonally and via feed-back inhibition and feed-forward activation.

Thus mobilization of free fatty acids from the adipose tissue results from low insulin levels.

The free fatty acids are then transported through the blood to the rest of the body including the liver.

In the liver fatty acid oxidation and ketone body synthesis is activated by glucagon.

Note that glucagon and insulin levels are opposite: high insulin =low glucagon and vice-versa.

So for low insulin will also have high glucagon, thus fatty acids will be released from the adipose and will be converted in the liver into ketone bodies.

The regulation of fatty acid oxidation, fatty acid synthesis and ketone body synthesis in the liver is summarized in the figure:

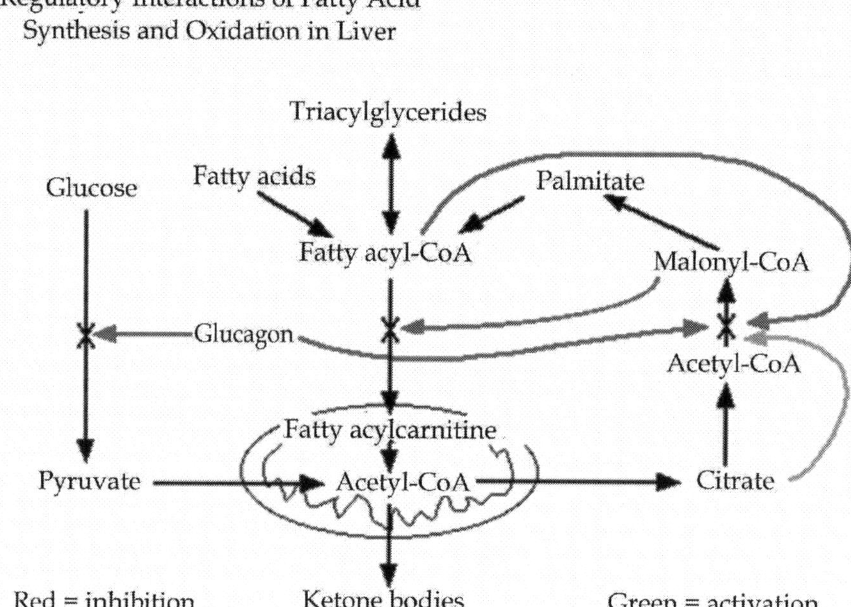

Note that a lack of insulin results in a release of fatty acids from adipose.

SYNTHESIS OF TRIACYLGLYCEROLS AND GLYCEROPHOSPHOLIPIDS

Triacylglycerol Biosynthesis (Esterification)

Formation of phosphatidic acid from glycerol or dihydroxyacetone phosphateand its conversion to triacylglycerol or phospholipids.

Most fatty acids are esterified as triacylglycerols or glycerophospholipids. Intermediate molecule in synthesis of these two molecules is phosphatidic acid or phosphatidate.

There are two pathways:

1) *de novo* – "from scratch"
2) **salvage pathway** - uses "old" pieces and parts to make new molecules

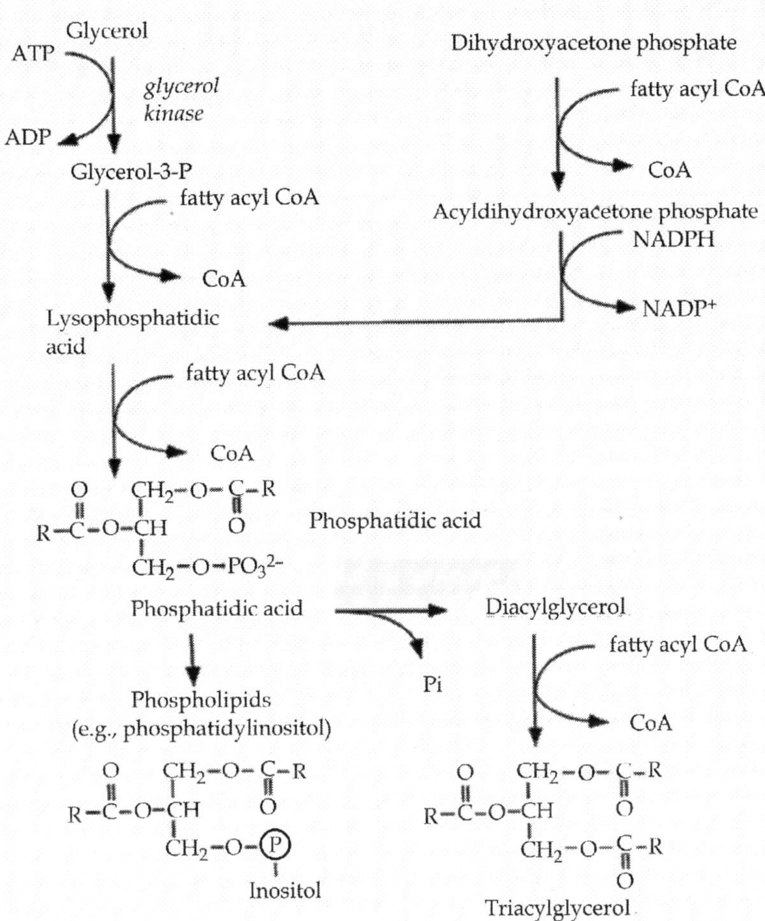

Synthesis of phosphatidate:

- Common intermediate in synthesis of phosphoglycerides and triacylglycerols
- Formed from glycerol 3-phosphate and 2 acetyl CoA molecules
- Enzyme is **glycerol phosphate acyltransferase**

Synthesis of triacylglycerols and neutral phospholipids:

- Uses phosphatidate, which is dephosphorylated to produce 1,2-diacylglycerol

  If acetylated → triacylglyerol

  If reacted with nucleotide derivative → phosphatidylcholine or phosphatidylethanolamine

Synthesis of acidic phospholipids:

- Uses phosphatidate and reacts it with CTP → CDP-diacylglycerol
  - Addition of serine → phosphatidylserine
  - Addition of inositol → phosphatidylinositol
- In mammals, phosphatidylserine and phosphatidylethanolamine can be interconverted - base-exchange occurs in ER.
- Decarboxylation occurs in mitochondria and procaryotes

Synthesis of Sphingolipids

- All have C18 unsaturated alcohol (sphingosine) as structural backbone, rather than glycerol
- Palmitoyl CoA and serine condense → dehydrosphinganine → sphingosine
- Acetylation of amino group of sphingosine → ceramide
- Substitution of terminal hydroxyl group gives:
  - sphingomyelin – addition of phosphatidylcholine
  - cerebroside – substitute UDP-glucose or UDP-galactose
  - gangliosides – substitute oligosaccharide

Tay-Sachs disease = inherited disorder of ganglioside breakdown.

- Deficient or missing enzyme is b-N-acetylhexosaminidase, which removes the terminal N-acetylgalactosamine residue from its ganglioside.
- One in 30 Jewish Americans of eastern European descent are carriers of a defective allele.
- Can be diagnosed during fetal development by assaying amniotic fluid for enzyme activity.
- Causes weakness, retarded psychomotor development, blindness by age two, and death around age three.

Synthesis of Cholesterol

- Precursor of steroid hormones and bile salts.
- Most cholesterol is synthesized in liver cells, although most animal cells can synthesize it.

- Starts with 3 molecules of acetyl CoA to form 3-hydroxy-3-methyl-glutaryl CoA, which is reduced to mevalonate (C6) by **HMG-CoA reductase** (first committed step of cholesterol synthesis)
- Amount of cholesterol formation by liver and intestine is highly responsive to cellular levels of cholesterol.
- Enzyme HMG-CoA reductase is controlled in multiple ways:
  1) Rate of enzyme synthesis is controlled by sterol regulatory element (SRE); SRE inhibits mRNA production
  2) Translation of reductase mRNA is inhibited by nonsterol metabolites derived from mevalonate
  3) Degradation of the enzyme occurs at high enzyme levels
  4) Phosphorylation of enzyme

  If enzyme is phosphorylated via glucagon pathway → decreased activity → cholesterol synthesis ceases when ATP levels are low

  If enzyme is dephosphorylated via insulin pathway → increased activity
- Cells outside liver and intestine obtain cholesterol from blood instead of synthesizing it *de novo*.
- Steps in the uptake of cholesterol by LDL pathway:
  1) apolipoprotein on surface of LDL particle binds to receptor on membrane of nonhepatic cells
  2) LDL-receptor complex internalized by endocytosis
  3) vesicles formed fuse with lysosomes, which breaks apart protein part of lipoprotein to amino acids and hydrolyzes cholesterol esters
  4) released unesterified cholesterol can be used for membrane biosynthesis or be reesterified for storage
- Defects in LDL receptor lead to **familial hypercholesterolemia** (FH), in which cholesterol and LDL levels are markedly elevated.
- Result is deposition of cholesterol in tissues because of high levels of LDL-cholesterol in blood
- Heterozygotes suffer from atherosclerosis and increased risk of stroke
- Homozygotes usually die in childhood from coronary artery disease
- Disease is the result of an absence (homozygotes) or reduction (heterozygotes) in number of LDL receptors.

- LDL entry into liver and other cells is impaired.
- Drug therapy can help heterozygotes
  1) can inhibit intestinal absorption of bile salts (which promote absorption of dietary cholesterol)
  2) lovastatin - competitive inhibitor of HMG-CoA reductase → blocks cholesterol synthesis

INTERRELATIONSHIP OF FAT AND CARBOHYDRATE (CHO) METABOLISM WHEN GLUCOSE IS HIGH:

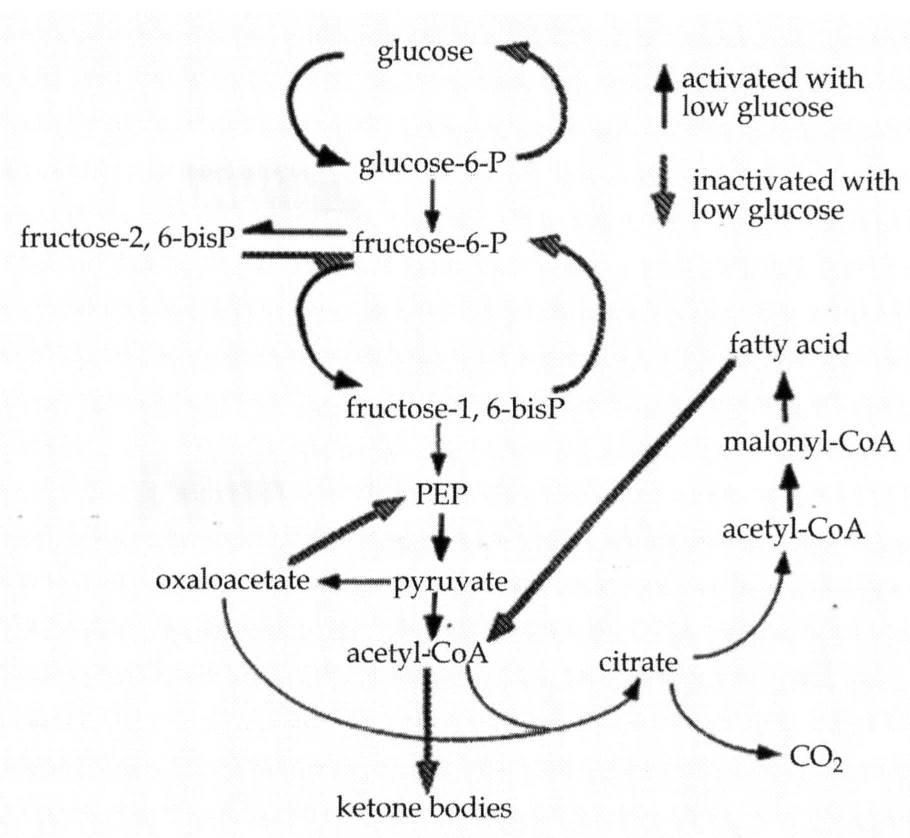

Conditions Favoring Fat Synthesis (*Lipogenesis*):

CHO intake → Elevated Blood Glucose → Insulin High, Glucagon Low

Insulin (-) hormone sensitive lipase; (+) glucose utilization (glycolysis) and acetyl CoA production for lipogenesis

Citrate (+) acetyl CoA carboxylase

Malonyl CoA (-) carnitine palmitoyl transferase I (decreasing ·-oxidation)

Conditions Favoring Lipolysis / b-*Oxidation*:

Starvation → Low Blood Glucose → Insulin Low, Glucagon High → (+) Lipolysis (free fatty acids for liver)

"Fight or Flight" → Epinephrine High → (+) Lipolysis (energy for muscle)

Low Blood Glucose → (+) gluconeogenesis (decreasing "C" supply for lipogenesis)

Acetyl CoA Carboxylase- phosphorylated ("Inactive")

CPT I- "Active" (due to decrease in Malonyl CoA)

Interrelationship of Fat and CHO Metabolism When Glucose is Low:

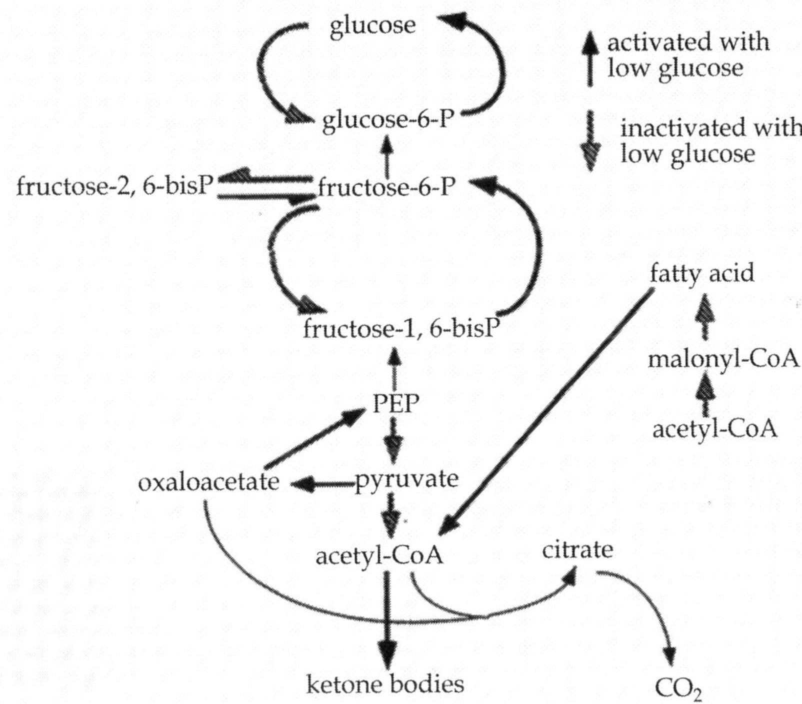

Starvation:

Fatty Acids are Oxidized in Liver which Promotes Gluconeogenesis by:

(a) Providing energy

(b) Generating NADH

(c) Forming acetyl CoA to activate pyruvate carboxylase

(d) Producing citrate which increases F-1,6 bisPase activity

Comparison of Energy Yields and Oxygen Consumption:

1 NADH = 3 ATP

1 $FADH_2$ = 2 ATP

Palmitate (3 molecules = 48 "C"s)

Step 1: β-oxidation to acetyl CoA (6 cycles); (1 NADH + 1 ADH2)/cycle

Step 1cont'd: 7 x 3 molecules 21 FADH2+ 21 NADH → +105 ATP; - 42 O atoms

Step 2: Acetyl CoA oxidation via TCA cycle; (3 NADH, 1 FADH2, 1 GTP) / Acetyl CoA

Step 2 cont'd: 8(6 + 2) x 3 molecules: 24 Acetyl CoA → +288 ATP; -96 O atoms

Step 3: Acyl CoA formation (ATP → AMP + PPi; 2 ATP / molecule)

Step 3 cont'd: 2 x 3 molecules: → - 6 ATP

α-hydroxybutyrate (12 molecules = 48 "C"s)

Step 1: oxidation to acetoacetate via b-hydroxybutyrate DH

Step 1 cont'd: 1 x 12 molecules: 12 NADH → +36 ATP; -12 O atoms

Step 2: Acetoacetate cleaved to 2 acetyl CoA (loss 1 GTP due to succinyl CoA diversion)

Step 2 cont'd: 1 x 12 molecules: -12 GTP → -12 ATP

Step 3: Oxidation of acetyl CoA via TCA cycle

Step 3 cont'd: 2 x 12 molecules: 24 Acetyl CoA → +288 ATP; -96 O atoms

Glucose (8 molecules = 48 "C"s)

Step 1: Aerobic glycolysis, 2 NADH (mal-asp shuttle) + 2 ATP/glucose

Step 1 cont'd: 2 x 8 molecules: 16 NADH + 16 ATP → +64 ATP; - 16 O atoms

Step2: PDH

Step 2 cont'd: 2 x 8 molecules: 16 NADH → +48 ATP; -16 O atoms

Step 3: Oxidation of acetyl CoA via TCA cycle

Step 3 cont'd: 2 x 8 molecules: 16 Acetyl CoA → +192 ATP; - 64 O atoms

Fuel	Total ATP	O_2 Used	ATP/"C"	CO_2/O_2	ATP/"O" Atom
Palmitate	387	69	8.12	0.7	2.80
Ketone	312	54	6.50	0.9	2.89
Glucose	304	48	6.33	1.0	3.17

Note: ATP / "O" Atom is a measure of "fuel" efficiency

Fats: Produce more ATP per "C", however it is at the expense of more oxygen(ATP/ O atom), thus the respiratory quotient (CO_2 / O_2) is the lowest.

Glucose: is the better fuel under conditions of O_2 limitation, giving more ATP / O atom.

Ketones: in starvation increase the amount of O_2 needed to burn fuel to CO_2 (CO_2 / O_2) only a small amount as compared to fats. The energy yield per carbon for ketones is similar to glucose. The brains energy needs and O_2 availability can be met nearly as well by ketones, as by glucose.

C. AMINO ACID METABOLISM

Will be interested in two things:

1) origin of nitrogen atoms and their incorporation into amino group
2) origin of carbon skeletons

Amino Acid Synthesis

Nitrogen fixation

Gaseous nitrogen is chemically unreactive due to strong triple bond.

To reduce nitrogen gas to ammonia takes a strong enzyme → reaction is called nitrogen fixation.

Only a few organisms are capable of fixing nitrogen and assembling amino acids from that.

Higher organisms cannot form NH_4^+ from atmospheric N_2.

Bacteria and blue-green algae (photosynthetic procaryotes) can because they possess **nitrogenase**.

Enzyme has two subunits:

1) strong reductase - has Fe-S cluster that supplies e- to second subunit
2) two re-dox centers, one of which is a nitrogenase

   Composed of iron and molybdenum that reduces N_2 to NH_4^+

Reaction is ATP-dependent, but unstable in the presence of oxygen.

Enzyme is present in *Rhizobium*, symbiotic bacterium in roots of legumes (i.e. soybeans)

Nodules are pink inside due to presence of **leghemoglobin** (legume hemoglobin) that binds to oxygen to keep environment around enzyme low in oxygen (nitrogen fixation requires the absence of oxygen)

Plants and microorganisms can obtain NH_3 by reducing nitrate (NO_3^-) and nitrite (NO_2^-) → used to make amino acids, nucleotides, phospholipids.

Assimilation of Ammonia

Assimilation into amino acids occurs through glutamate and glutamine.

α-amino group of most amino acids comes from a-amino group of glutamate by transamination.

Glutamine contributes its side-chain nitrogen in other biosynthetic reactions.

Reaction:

$$NH_4^+ + \alpha\text{-ketoglutarate} \underset{\text{glutamate dehydrogenase}}{\overset{NADPH + H^+ \quad NADP^+}{\rightleftharpoons}} \text{glutamate} + H_2O$$

Another reaction that occurs in some animals is the incorporation of ammonia into glutamine via **glutamine synthetase**:

$$\text{glutamate} + NH_4^+ + ATP \longrightarrow \text{glutamine} + ADP + P_i + H^+$$

When ammonium ion is limiting, most of glutamate is made by action of both enzymes to produce the following (sum of both reactions):

$$NH_4^+ + \alpha\text{-ketoglutarate} + NADPH + ATP \longrightarrow \text{glutamate} + NADP^+ + ADP + P_i$$

Transamination Reactions

Having assimilated the ammonia, synthesis of nearly all amino acids is done via **tranamination** reactions.

Glutamate is a key intermediate in amino acid metabolism

Amino group is transferred to produce the corresponding amino acid.

$$\text{amino acid}_1 + \text{keto acid2} \overset{\textbf{transaminase}}{\longleftrightarrow} \text{keto acid}_1 + \text{amino acid}_2$$

Origins of Carbon Skeletons of the Amino Acids

Amino acids that must be supplied in diet are termed **essential**; others are **nonessential**.

Although the biosynthesis of specific amino acids is diverse, they all share a common feature - carbon skeletons come from intermediates of glycolysis, PPP, or citric acid cycle.

There are only six biosynthetic families:

1) Derived from oxaloacetate → Asp, Asn, Met, Thr, Ile, Lys

2) Drived from pyruvate → Ala, Val, Leu
3) Derived from ribose 5-phosphate → His
4) Derived from PEP and erythrose 4-phosphate → Phe, Tyr, Trp
5) Derived from a-ketoglutarate → Glu, Gln, Pro, Arg
6) Derived from 3-phosphoglycerate → Ser, Cys, Gly

Porphyrin Synthesis

First step in biosynthesis of porphyrins is condensation of glycine and succinyl CoA to form a-aminolevulinate via a-aminolevulinate synthase.

Translation of mRNA of this enzyme is feedback-inhibited by heme

Second step involves condensation of two molecules of a-aminolevulinate to form porphobilinogen; catalyzed by a-aminolevulinate dehydrase.

Third step involves condensation of four porphobilinogens to form a linear tetrapyrrole via porphobilinogen deaminase.

This is cyclized to form uroporphyrinogen III.

Subsequent reactions alter side chains and degree of saturation of porphyrin ring to form protoporphyrin IX.

Association of iron atom creates heme; iron atom transported in blood by transferrin.

Inherited or acquired disorders called **porphyrias** are result of deficiency in an enzyme in heme biosynthetic pathway.

congenital erythropoietic porphyria - insufficient cosynthase (cyclizes tetrapyrrole)

Lots of uroporphyrinogen I, a useless isomer are made

RBCs prematurely destroyed

Patient's urine is red because of excretion of uroporphyrin I

Heme Degradation

Old RBCs are removed from circulation and degraded by spleen.

Apoprotein part of hemoglobin is hydrolyzed into amino acids.

First step in degradation of heme group is cleavage of ?-methene bridge to form biliverdin, a linear tetrapyrrole; catalyzed by heme oxygenase; methene bridge released as CO.

Second step involved reduction of central methene bridge to form bilirubin; catalyzed by biliverdin reductase.

Bilirubin is complexed with serum albumin ® liver ® sugar residues added to propionate side chains.

2 glucuronates attached to bilirubin are secreted in bile.

Jaundice - yellow pigmentation in sclera of eye and in skin —> excessive bilirubin levels in blood

Caused by excessive breakdown of RBCs, impaired liver function, mechanical obstruction of bile duct.

Common in newborns as fetal hemoglobin is broken down and replaced by adult hemoglobin.

Amino Acid Catabolism

Excess amino acids (those not used for protein synthesis or synthesis of other macromolecules) cannot be stored.

Surplus amino acids are used as metabolic fuel.

α-amino group is removed; carbon skeleton is converted into major metabolic intermediate

Amino group converted to urea; carbon skeletons converted into acetyl CoA, acetoacetyl CoA, pyruvate, or citric acid intermediate.

Fatty acids, ketone bodies, and glucose can be formed from amino acids.

Major site of amino acid degradation is the liver.

First step is the transfer of amino group to ketoglutarate to form glutamate, which is oxidatively deaminated to yield $NH4^+$ (see pathway sheet).

Some of $NH4^+$ is consumed in biosynthesis of nitrogen compounds; most terrestrial vertebrates convert $NH4^+$ into urea, which is then excreted (considered **ureotelic**).

Terrestrial reptiles and birds convert $NH4^+$ into uric acid for excretion (considered **uricotelic**).

Aquatic animals excrete $NH4^+$ (considered **ammontelic**).

In terrestrial vertebrates $NH4^+$ is converted to urea via **urea cycle**.

One of nitrogen atoms in urea is transferred from aspartate; other is derived from NH_4^+; carbon atom comes from CO_2.

Urea Cycle

There are six steps of the urea cycle:

1) Bicarbonate ion, NH_4^+ and 2 ATP necessary to form carbamoyl phosphate via carbamoyl phosphate synthetase I (found in mitochondrial matrix).

2) Carbamoyl phosphate and ornithine (carrier or carbon and nitrogen atoms; an amino acid, but not a building block of proteins) combine to form citrulline via ornithine transcarbamoylase

3) Citruilline is transported out of mitochondrial matrix in exchange for ornithine

4) Citruilline condenses with aspartate → arginosuccinate via an ATP-dependent reaction via arginosuccinate synthetase

5) Arginosuccinate cleaved to form fumarate and arginine via arginosuccinate lyase

   fumarate → malate → oxaloacetate → gluconeogenesis

   oxaloacetate has four possible fates:

   1) transamination to aspartate
   2) conversion into glucose via gluconeogenesis
   3) condensation with acetyl CoA to form citrate
   4) conversion into pyruvate

6) Two -NH_2 groups and terminal carbon of arginine cleaved to form ornithine and urea via arginase

   Ornithine is transported into mitochondrion to repeat cycle

Overall reaction:

$CO_2 + NH_4^+ + 3\,ATP + aspartate + 2\,H_2O \rightarrow urea + 2\,ADP + 2\,P_i + AMP + PP_i + fumarate$

Inherited defects in urea cycle:

1) Blockage of carbamoyl phosphate synthesis leads to hyperammonemia (elevated levels of ammonia in blood)

2) argininosuccinase deficiency

   Providing surplus of arginine in diet and restricting total protein intake

   Nitrogen is excreted in the form of argininosuccinate

3) carbamoyl phosphate synthetase deficiency or ornithine transcarbamoylase deficiency

   Excess nitrogen accumulates in glycine and glutamine; must then get rid of these amino acids

Done by supplementation with benzoate and phenylacetate (both substitute for urea in the disposal of nitrogen)

benzoate → benzoyl CoA → hippurate

phenylacetate → phenylacetyl CoA → phenylacetylglutamine

Fate of Carbon Skeleton of Amino Acids

Used to form major metabolic intermediates that can be converted into glucose or oxidized by citric acid cycle.

All 20 amino acids are funneled into seven molecules:

1) pyruvate
2) acetyl CoA
3) acetoacetyl CoA
4) α-ketoglutarate
5) succinyl CoA
6) fumarate
7) oxaloacetate

Those that are degraded to acetyl CoA or acetoacetyl Coa are termed **ketogenic** because they give rise to ketone bodies.

Those that are degraded to pyruvate or citric acid cycle intermediates are termed **glucogenic**.

Leucine and lysine are only ketogenic → cannot be converted to glucose

Isoleucine, phenylalanine, tryptophan, tyrosine are both.

All others are glucogenic only.

C_3 family (alanine, serine, cysteine) → pyruvate

C_4 family(aspartate and asparagine) → oxaloacetate

C_5 family (glutamine, proline, arginine, histidine) → glutamate → a-ketoglutarate

Methionine, isoleucine, valine, threonine → succinyl CoA

Leucine → acetyl CoA and acetoacetate

Phenylalanine and tyrosine → acetoacetate and fumarate

Tryptophan → pyruvate

Regulation of the Urea Cycle

The main allosteric enzyme is **glutamate dehydrogenase**.

It is inhibited by high GTP and ATP levels.

It is stimulated by high GDP and ADP levels.

Protein Catabolism

Peptidases

Exopeptidases

Exopeptidases hydrolyse the polypeptides either from the carboxyl terminus (carboxypeptidases) or from the amino terminus (amino peptidase). Some of the important exopeptidases and their sources are given in below

Exopeptidase	Source
Carboxypeptidase A	Bovine pancreas
Carboxypeptidase B	Bovine Pancreas
Carboxypeptidase C	Citrus leaves
Amino peptidase M	Porcine kidney
Leucine amino peptidase	Porcine kidney

Endopeptidases

Enzymes, which hydrolyse the internal peptide bonds, are called as endopeptidases. Endopeptidases have side chain requirements for the residues flanking the scissile peptide bond. (Peptide bond to be cleaved). The important endopeptidases and their sources are listed below

Enzyme	Source
Trypsin	Bovine pancreas
Chymotrypsin	Bovine pancreas
Pepsin	Bovine gastric
Elastase	Bovine pancreas
Papain	Papaya latex

Amino Acid Metabolism

The degradation of the carbon skeletons of 20 amino acids converges to just seven metabolic intermediates namely.

- Pyruvate
- Acetyl CoA
- Succinyl CoA
- Fumarate

- Acetoacetyl CoA
- Oxaloacetate
- α-Ketoglutarate

Catabolism of amino acids

Nitrogen can be eliminated in a variety of forms, depending on the physiological conditions experienced by the organism:

- Ammonia and/or ammonium ion.
- Urea. Animals with readily available water (for drinking) use urea as their major nitrogenous waste product. Urea is non-toxic at low (<1 M) concentrations and is extremely soluble (>10 M).
- Uric acid. Animals living in deserts, environmental or physiological, use uric acid. Uric acid is thus used by reptiles and birds, including species living in or near water.

The product ammonia is excreted after conversion to urea or other products and the carbon skeleton is degraded to CO_2 releasing energy. The important reaction involved in the deamination of amino acids is

- Transamination
- Oxidative deamination
- Non oxidative deamination

D. NUCLEOTIDE METABOLISM

Roles of nucleotides in the cell:

1) Activated precursors of DNA and RNA
2) Nucleotide derivatives are activated intermediates in many biosynthetic pathways

   e.g. UDP-glucose, CDP-diacylglycerol
3) Universal currency of cell (i.e. ATP)
4) Components of three major coenzymes: NAD+, FAD, and CoA
5) Metabolic regulators (i.e. cyclic AMP)

   Nucleotide synthesis can either by de novo or by recycling preformed bases (salvage pathway).

Nomenclature:

Nucleotides are composed of three components:

1) nitrogenous base - pyrimidine (cytosine, uracil, thymine) or purine (adenine or guanine)
2) pentose sugar - ribose (RNA) or deoxyribose (DNA)
3) phosphate group

nucleoside - purine or pyrimidine base linked to pentose sugar

nucleotide - phosphate ester of nucleoside

SYNTHESIS OF PURINE NUCLEOTIDES

- Purine ring is synthesized de novo from 5 different precursors: aspartate (N-1 atom), CO_2 (C-6 atom), glycine (C-4, C-5, N-7 atoms), tetrahydrofolate (C-2, C-8 atoms) and glutamine (N-3, N-9).
- Purine ring structure is synthesized from ribose 5-phosphate; PRPP then donates ribose 5-phosphate for purine synthesis.
- Purine ring is built onto the ribose 5-phosphate via a 10 -tep pathway: glutamine, glycine, tetrahydrofolate, and glutamine make contributions to form 5-membered ring; construction of 6-membered ring forms inosine 5'-monophosphate (IMP).
- IMP can be converted into AMP or GMP
  - For AMP synthesis, aspartate amino group condenses with keto-group of IMP; GTP-dependent reaction
  - For GMP synthesis, C-2 is oxidized to form xanthosine monophosphate (XMP)

    Amide nitrogen of glutamine replaces oxygen of C-2 to form GMP

    ATP-dependent reaction

Synthesis of purine bases using the salvage pathway:

- Free purine bases are formed by degradation of nucleic acids and nucleotides.
- Purine nucleotides can be synthesized from preformed bases by salvage reactions (simpler and less costly than de novo pathway).
- Ribose phosphate portion of PRPP is transferred to purine to form the corresponding ribonucleotide:

purine PPi

PRPP ⟶ purine nucleotide

Two salvage enzymes recover purine bases:

1) adenine phosphoribosyl transferase

Adenine + PRPP ⟶ Adenylate + PPi

2) hypoxanthine-guanine phosphoribosyl transferase (HGPRTase)

Hypoxanthine + PRPP ⟶ Inosinate + PPi

Guanine + PRPP ⟶ Guanylate + PPi

Regulation of purine nucleotide synthesis:

Probably largely by feedback inhibition.

Glutamine-PRPP amidotransferase (in main pathway) is allosterically inhbited by IMP, AMP, GMP.

Those steps leading specifically to AMP or GMP synthesis work primarily by feedback inhibition

XMP and GMP inhibit IMP dehydrogenase

AMP inhibits adenylosuccinate synthetase

SYNTHESIS OF PYRIMIDINE NUCLEOTIDES

- Pyrimidine ring is assembled first, then linked to ribose phosphate → pyrimidine nucleotide.
- Requires fewer ATPs than purine synthesis (2 vs. 4).
- Pyrimidine ring has three metabolic precursors: bicarbonate, amide group of glutamine, aspartate.
- PRPP is also required.
- There is a 6-step pathway for de novo synthesis of UMP:
  1) Glutamine combines with bicarbonate ion + 2ATPs to yield carbamoyl phosphate + glutamate
  2) Carbamoyl phosphate combines with aspartate via aspartate transcarbamolyase to form carbamoyl aspartate (product contains all the atoms necessary for pyrimidine ring).
  3) Arbamoyl phosphate is cyclized enzymatically to form L-dihydroorotate.

4) L-dihydroorotate is oxidized by dihydroorotate dehydrogenase to form orotate; e- removed from substrate are transferred to ubiquinone → O2 to ETS.

5) Orotate replaces pyrophosphate group of PRPP to form orotidine 5'-monophosphate(OMP)

6) OMP is decarboxylated by OMP decarboxylase to form uridine 5'-monophosphate (UMP)

Dihydroorotate is produced in the cytosol, then passes through the outer mitochondrial membrane.

Enzyme dihydroorotate DH is on outer surface of inner mitochondrial membrane

Orotate then moves back into cytosol

Regulation of UMP synthesis:

asparate carbamoylase (ATCase)- main regulatory enzyme

Inhibited by UTP and CTP

Activated by ATP

Keeps purines and pyrimidines in equal amounts

Synthesis of CTP

Formation of CTP from UMP in three reactions (see pathway sheet).

Regulation of pathway is via CTP synthetase

- allosterically inhibited by CTP

CONVERSION OF RIBONUCLEOTIDES TO DEOXYRIBONUCLEOTIDES

Deoxyribonucleotides are formed from ribonucleotides by ribonucleoside diphosphate reductase.

Energy to fuel reduction comes from NADPH.

There are really three proteins involved:

1) thioredoxin reductase

2) thioredoxin

3) ribonucleotide reductase

Once dADP, dGDP, and dCDP are formed, they are phosphorylated by nucleoside diphosphate kinases.

Regulation of ribonucleoside diphosphate reductase is complex because there are 2 regulatory sites:

1) Activity site - a.k.a. allosteric site - controls catalytic site
2) Specificity site - also allosterically regulated- controls substrate specificity

If ATP is bound in activity, enzyme is ACTIVE

If dATP or ATP is bound, reductase is pyrimidine specific

CDP → dCDP

UDP → dUDP

Binding of dTTP to specificity site causes enzyme to take GDP --> dGDP.

Binding of dGTP to specificity site causes enzyme to take ADP --> dADP.

Synthesis of Deoxythymidylate (dTMP) by Methylation of dUMP

dTMP is formed from dUMP, which is formed by any of the following:

dUDP + ADP → dUMP + ATP enzyme is nucleoside **monophosphate kinase**

dUDP + ATP ⟶ dUTP ⟶ dUMP + PPi

dCMP + H_2O ⟶ dUMP + NH4+

dUMP is converted to dTMP by thymidylate synthase

Methyl group donor is methylene tetrahydrofolate

Many cancer drugs inhibit the activity of thymidylate synthase and dihydrofolate reductase → decreased levels of dTMP synthesis → decreased DNA synthesis

SALVAGE OF PURINES AND PYRIMIDINES

Purine Catabolism

Many organisms convert purine nucleotides to uric acid

AMP → IMP → hypoxanthine →

GMP → xanthine →

Lesch-Nyhan syndrome

Total lack of HGPRTase.

Results in compulsive self-destructive behavior.

Self-mutilation, mental deficiency, spasticity.

Elevated levels of PRPP → increased rate of purine biosynthesis by de novo pathway → overproduction of uric acid.

Possible that brain may rely heavily on salvage pathway for IMP and GMP synthesis.

Shows that abnormal behavior can be caused by absence of a single enzyme.

Pyrimidine Catabolism

Begins with the hydrolysis of nucleosides and Pi from nucleotides.

Successive reactions produce ribose 1-phosphate or deoxyribose 1-phosphate.

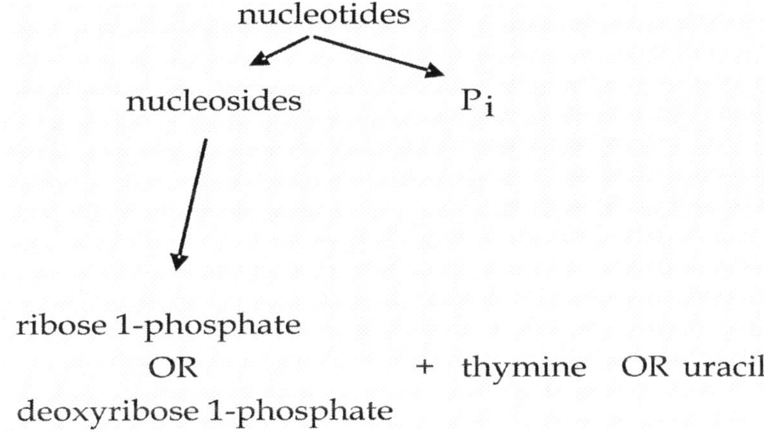

Thymine is ultimately broken down to succinyl CoA.

Uracil and cytosine are broken down into alanine, then acetyl CoA.

CHAPTER - 11

Photosynthesis

MOST OF THE ENERGY FOR LIFE COMES FROM SUNLIGHT

Total rate of energy delivery from sun to earth (at the top of the atmosphere) = 175,000 terrawatts

Terrawatt = 10^12 watts

Rate of energy capture by plants from the sun = net primary productivity (NPP) = 100 terrawatts

Only ~ 0.06% of sunlight is captured

Captured in chemical bonds, especially in sugar

Small amounts of energy are available to specialized microorganisms from chemical reactions not linked to sunlight

It is estimated that 99% of the energy used by living cells comes from the sun

Incorporation of sunlight into chemical bonds occurs through the process of photosynthesis

"Invented" by cyanobacteria about 2 billion years ago

Plants, Cyanobacteria and Algae are Autotrophs

Autotrophs make their own macromolecules and do not require products from other living creatures for life

Other types of organisms, including ourselves, are heterotrophs: require products from other species for energy and materials

Photosynthesis

Photosynthesis is the process of converting light energy to chemical energy and storing it in the bonds of sugar. This process occurs in plants and some algae (Kingdom Protista). Plants need only light energy, CO2, and H2O to make sugar. The process of photosynthesis takes place in the chloroplasts, specifically using chlorophyll, the green pigment involved in photosynthesis.

In photosynthesis, carbon dioxide and water are combined, using energy from sunlight, to form glucose

Oxygen is given off as a waste product

This is the source of the oxygen in the atmosphere

Photosynthesis has 2 sets of reactions

Light reactions - split water

Hydrogens are used to produce a reduced coenzyme, NADPH, and ATP

Oxygen is given off

Once the NADPH and ATP are formed, the rest of the reactions can take place in the dark

CO2 is reduced to glucose

A set of cyclic reactions: the Calvin cycle

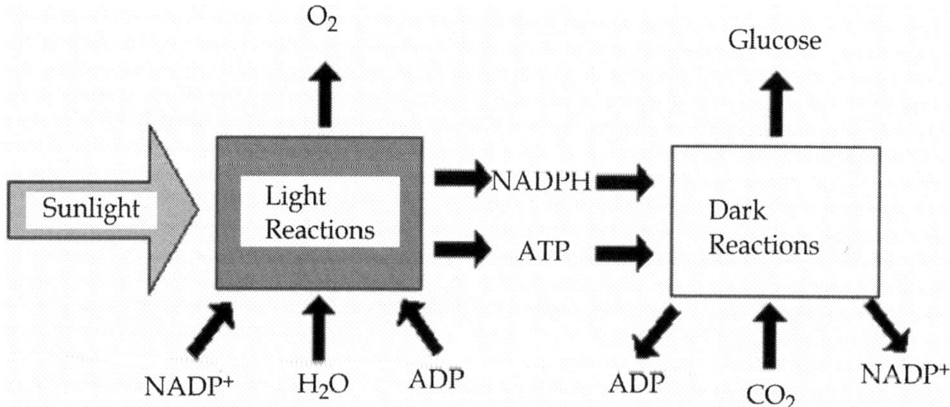

Energy budget of photosynthesis to produce one molecule of glucose

Light reactions:

$$12H_2O + 12NADP + 18ADP \rightarrow 6O_2 + 12NADPH + 18ATP$$

Dark reactions

$$6CO_2 + 12NADPH + 18ATP \rightarrow C_6H_{12}O_6 + 12NADP + 18ADP + 6H_2O$$

LIGHT REACTION OF PHOTOSYNTHESIS

Photosynthesis Starts with the Absorption of Light

Leaves can absorb 90% of the light striking them

Absorption is by a number of pigments including:

Chlorophyll a		Main photosynthetic pigment
Chlorophyll b		Accessory pigment: passes light to chlorophyll a
Xanthophylls		More accessory pigments
Carotene		Another accessory pigment

Chlorophyll has a light-sensitive ring with a Mg atom in the center. It is anchored in chloroplast membranes by a hydrophobic tail

Plants have chlorophylls a & b; algae also have 2 other types (c & d: see table, p. 533 of text)

Chlorophyll absorbs mostly light in the blue and red regions of the spectrum; it appears green because it does not absorb green light

Light absorption is affected by the environment

Chlorophyll a has an absorption peak either at 680 or 700 nm depending upon what proteins it is associated with (photosystems I & II, see below)

The use of accessory pigments allows photosythesis to use a large proportion of the visible light

Once light is absorbed several things can happen. Suppose violet light is absorbed:

Violet light can be re-emmitted

The light energy can be converted to heat

Some of the energy can be given off as heat and the rest as light; the light will have a lower energy, which gives it a different color, such as red (fluorescence):

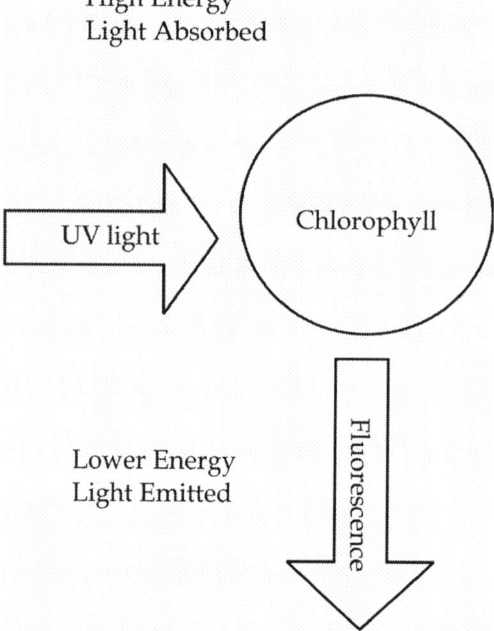

If you have an organized photocenter the light energy can be passed from molecule to molecule and some of it can be used to make chemical bonds

Photocenter channels light into photosynthesis, reducing fluorescence and other forms of re-emmission

Excitations by light do not last long, usually about 1 nanoseconds or less

Chlorophyll is Located in the Thylakoid Membranes of Chloroplasts

Chloroplasts are greenish in color, about 5 microns in length (many variations in shape)

Originated (probably) by endosymbiosis

More complicated structure than mitochondria: 3 compartments (Compare mitochondria & chloroplasts.):

Have hollow stacked compartments

Compartments are called thylakoids

Stacks are called grana: about 10 compartments per granum

Stacks are interconnected

Thylakoids form by pinching off from chloroplast inner membrane

Thylakoid membranes have the photosynthetic apparatus

Chlorophyll and accessory pigments

Electron transport chain

Hydrogen pumps

ATP synthase

The Thylakoid Membranes of Chloroplasts Have 2 Photosystems

In the chloroplast membranes the light-capturing pigments are organized into photosystems:

Accessory pigments (about 300-400 molecules) clustered around a reaction center containing chlorophyll a

Proteins hold system together and catalyze some steps

There are two types of photosystem

Photosystem I:

Reaction center absorbs most efficiently at 700 nm (is called P700)

Passes electrons through acceptors that generate NADPH

Photosystem II:

Reaction center absorbs most efficiently at 680 nm (is called P680)

Passes electrons through acceptors that generate ATP

Splits water, making oxygen

Photosystems work together:

PSII comes 'earlier' in the reaction series than PSI.

Photosystem II feeds electrons to photosystem I

Emerson enhancement effect: photosynthesis requires light of 2 different wavelengths (around 680 and 700) at the same time

Each granum has about 200 of each of the two photosystems

Electron Flow and Hydrogen Gradients Generate ATP and NADPH in the Chloroplast

Photosystem II is excited first in the light reactions

The excited electron is passed to an electron transport system

To replace the electron lost from the reaction center an electron is pulled from water

This generates H+ ions and oxygen gas:

$2H_2O \rightarrow 4\ H+ + O_2 + 4e-$

As it is passed from one acceptor to another energy is released which is used to pump hydrogen ions from the stroma to the thylakoid compartment

The thylakoid compartment pH drops to around 5, compared to the stroma, which rises to about 8

The hydrogen gradient causes hydrogen ions to flow through the channels in ATP synthase and this generates ATP in the stroma

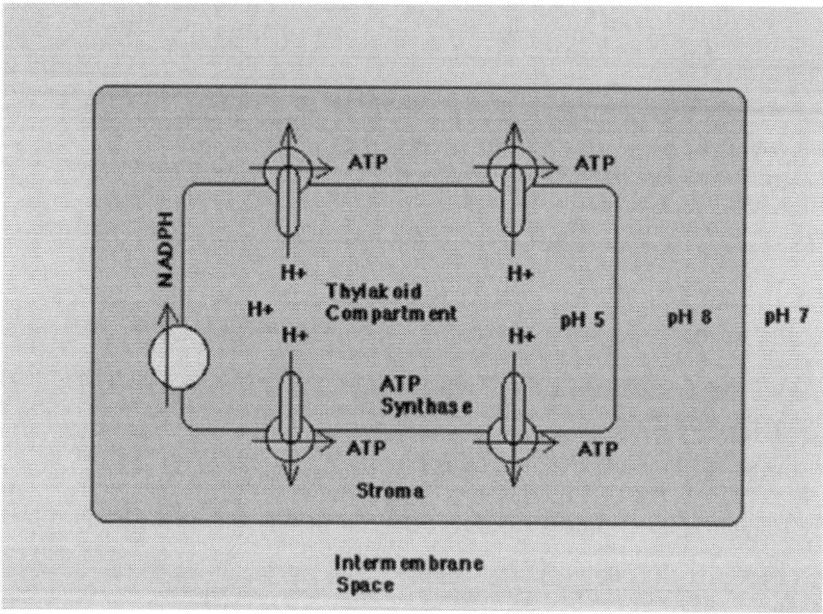

While the electron is still in a partially excited state it is passed to photosystem I

A second photon excites the electron

It is passed to other acceptors, finally to NADP, where it produces NADPH in the stroma

Final products of the light reactions are ATP and NADPH

The Light Reactions Split Water and Release Oxygen to the Atmosphere

Oxygen is a waste product of photosynthesis

All oxygen in atmosphere is believed to originate from photosynthesis

Photosynthesis began with cyanobacteria, about 3 billion years ago

First oxygen released reacted with iron, producing dark red bands in the rocks (Fe rust)

Atmospheric oxygen accumulation began after all Fe had reacted, approximately 2 billion years ago

Oxygen atmosphere created a crisis for life, very toxic substance

Some forms adapted, took advantage of extra energy available from oxygen chemistry

Oxygen atmosphere produced ozone layer

Protection against harmful UV light from sun

DARK REACTION OF PHOTOSYNTHESIS

The Dark Reactions are Driven by the Products of the Light Reactions

The light reactions supply vital materials to the dark reactions:

NADPH: supplie activated hydrogens to reduce carbon atoms

ATP: supplies energy to drive several steps

Making a single sugar molecule requires 18 ATPs and 12 NADPHs

Carbon Fixation is Catalyzed by Rubisco

In the first of the dark reactions CO_2 gas is added to a 5 carbon sugar, ribulose bisphosphate

This makes a 6 carbon intermediate, which immediately splits into two 3-phosphoglycerates

The reaction is catalyzed by ribulose bisphosphate carboxylase, "Rubisco", which is probably the most abundant enzyme on earth (it is 10-25% of leaf protein)

Calvin Cycle Reactions Make Sugar From 3-Phosphoglycerate

The diagram below is a simplified outline of the Calvin Cycle reactions which convert CO_2 to glucose and other sugars

The numbers indicate the numbers of molecules needed to eventually produce 1 glucose molecule

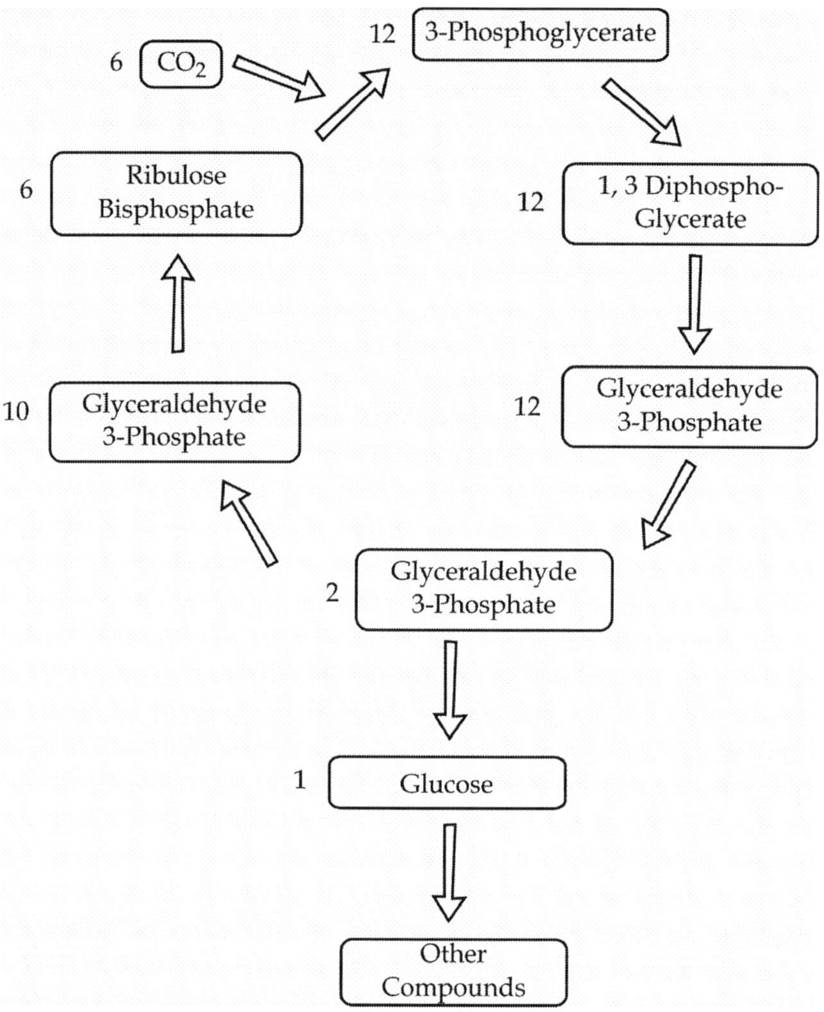

The 12 glyceraldehyde-3-phosphates formed are split into 2 groups:

10 are used to regenerate the ribulose bisphosphate and keep the cycle going

2 are condensed into a glucose molecule

The diagram does not show the details of the reactions used in regeneration and glucose synthesis (about 10 reactions)

Plants Convert Glucose to Sucrose or Starch

Plant cells have very little glucose

Most of the glucose is immediately converted to:

Starch in the chloroplasts and amyloplasts

Sucrose in the cytosol

Sucrose is Transported from the Leaf to the Rest of the Plant Through the Phloem

Vascular plants (flowering plants, gymnosperms, ferns, horsetails, lycopods, Psilotum) have vessels which conduct fluid through the plant.

Xylem: carries water and minerals from roots upward (transpiration)

Phloem: carries sugars and other nutrients from the leaves downward

Sucrose is loaded into the phloem by active transport.

Photosynthetic Gases Enter the Leaf Through Stomata

In higher plants photosynthesis usually takes place in the leaf

Leaves must take up CO_2 and give off O_2

Gases enter leaf through pores called stomata

Stomata can open and close

Surrounded by 2 balloon-shaped guard cells

When guard cells swell (by osmosis) the stomata open

When guard cells shrink stomata close

Stomata also give up water vapor

Some of this is necessary to pull water through the xylem

Can cause dehydration and wilting if the water is not replaced fast enough

Plants Living in Hot Environments Have Evolved Photosynthetic Variations that Minimize Water Loss

Plants in hot climates lose large amounts of water through their stomata when they are open

Must be open part of the time to pick up CO_2

Two alternative methods of photosynthesis have evolved to conserve water

Crassulacean acid metabolism (CAM): CO_2 is taken up at night when it is not as hot (less water loss)

The CO_2 taken up is bound to form 4 carbon acids

In the daytime the stomata close and the plant does photosynthesis with the supply of CO_2 taken up during the night

C4 Metabolism

Similar to CAM

Photorespiration

Process that reduces the yield of photosynthesis, because the active site of rubisco accepts O_2 in place of CO_2 and generates no ATP. This usually occurs on hot, dry days when stomata are closed and the O_2 concentration in the leaf exceeds that of CO_2, thereby competing for a common active site.

C4 metabolism

C4 photosynthesis is an adaptation to growth in hot climates. The first products of photosynthesis are C4 acids: 'normal' plants produce PGA, a 3 carbon acid, and hence are termed C3 plants. C4 photosynthesis has evolved independently in at least 16 families of flowering plants. C4 metabolism is a way of getting round the problem of Rubisco's unfortunate oxygenase activity. C4 plants concentrate CO_2 biochemically in a variety of ways, so Rubisco is exposed to a very high CO_2/O_2 ratio, which inhibits photorespiration.

Some of the most productive crops are C4, including maize and sugarcane; as are some of the worst weeds, such as *Amaranthus* (love-lies-bleeding) and *Tribulus terrestris*. Provided with a high light intensity, their maximum rates of photosynthesis may be double those of C3 plants.

C4 also metabolism reduces *transpiration* by lowering the CO_2 compensation point. Photosynthesis is slower when there is less CO_2 about, and the compensation point is defined as the $[CO_2]$ needed to give net photosynthesis.

C4 photosynthesis is found in both monocotyledons and dicotyledons. Most of the economically important ones are found in the Poaceae (grasses), and there are some interesting intermediates between C3 and C4, such as *Panicum milioides*. The genus *Atriplex* contains both C3 and C4 types which are still closely enough related to interbreed.

The C4 syndrome was discovery when radioactive $^{14}CO_2$ was fed to sugarcane, and the first products seen were C4 acids, not C3 PGA.

The receptor for CO_2 in C4 plants is not RuBP, but phosphoenolpyruvate (PEP), which is generally carboxylated by PEP carboxylase.

PEP + CO_2 + H_2O ® Oxaloacetate + P_1

CO_2 fixation into C4 acids occurs in the mesophyll. C4 acids are translocated into the bundle-sheath, through plasmodesmata, where they are decarboxylated to give CO_2, and this CO_2 enters the Calvin cycle as normal. The C_3 fragment is returned to the mesophyll and is metabolised back to PEP.

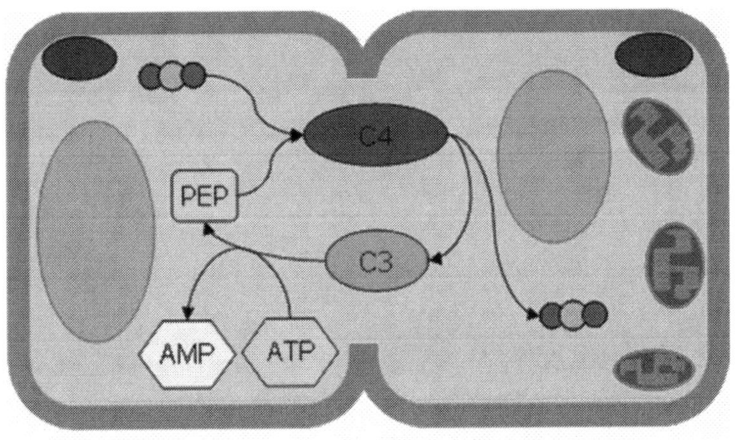

There are several advantages to C4 metabolism. Firstly CO_2 is released at a much higher concentration (about 10 times atmospheric) in the bundle sheath, where is competes better with oxygen for Rubisco (Rubisco is only found in the bundle sheath). Secondly, it allows net photosynthesis with less carbon dioxide, so the stomatal aperture may be held more nearly closed, and less water is lost. However, there are also disadvantages (or else almost all plants would run C4). The C_3 acids produced by the decarboxylation stage must be converted back to PEP. There are *three* ways of doing this, and they all use up ATP. This means C4 plants need higher light levels to generate the extra ATP. This helps to explain why they are found only in hot, sunny climates.

CAM metabolism

There is a second way of inhibiting photorespiration and transpiration, and that is using a process termed CAM. Whereas C4 metabolism physically separates CO_2 fixation and RuBP carboxylation - CO_2 is fixed in the mesophyll, and RuBP is carboxylated in the bundle-sheath - CAM (crassulacean acid metabolism) plants separate these processes *in time*. CAM is even better than C4 at inhibiting photorespiration, and is found in plants from really arid regions.

- Cacti.
- *Crassula.*
- Pineapple.

The biochemistry of CAM is very similar to C4. PEP carboxylase generates oxaloacetate; this is reduced to malate; and malate is split by NADP-malic enzyme to make pyruvate and CO_2. Hence photorespiration is inhibited. However, the anatomy is nothing special: there is no kranz anatomy; vacuoles in the mesophyll are often large, but not always. So how do they manage to inhibit photorespiration? The clue is in the fact that the pH of the vacuolar sap increases during the day and decreases at night. CAM plants synthesise acids (malate) during the night, and these are used up during the day. This is a temporal separation of fixation and RuBP carboxylation.

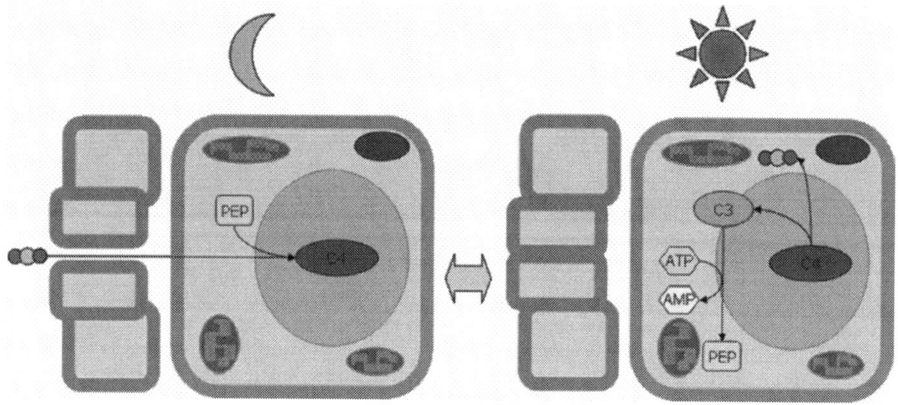

CO_2 is absorbed by PEP carboxylase during the night when the stomata are open. Malate is stored in the vacuole until morning and used to generate CO_2 during the day. The stomata can stay closed all day, increasing the internal $[CO_2]$, and this reduces photorespiration and water loss hugely.

CHAPTER - 12

Mitosis : Cell Reproduction, Growth and Repair

NEW CELLS ARE PRODUCED BY DIVISION OF PREVIOUSLY EXISTING CELLS

All cells come from previously existing cells

To make a new cell identical to itself a cell must perform 4 tasks:

- It must grow and make additional copies of all its organelles, enzymes, etc..
- It must duplicate its DNA
- It must accurately separate the DNA into 2 units, so that each cell gets a full set
- It must divide into 2 cells (cytokinesis)

These operations are called the cell cycle

- Duplication of material takes place in interphase (G1 + S + G2); about 24 hours in mammalian cells:
  - G1 = first gap
  - S = period of DNA synthesis (chromosomes duplicated)
  - G2 = second gap
  - Proteins and organelles duplicate in all parts of interphase (G1 + S + G2)

The last 2 operations are called mitosis (M in diagram); about 30 min in mammalian cells; the cycle goes clockwise in the diagram

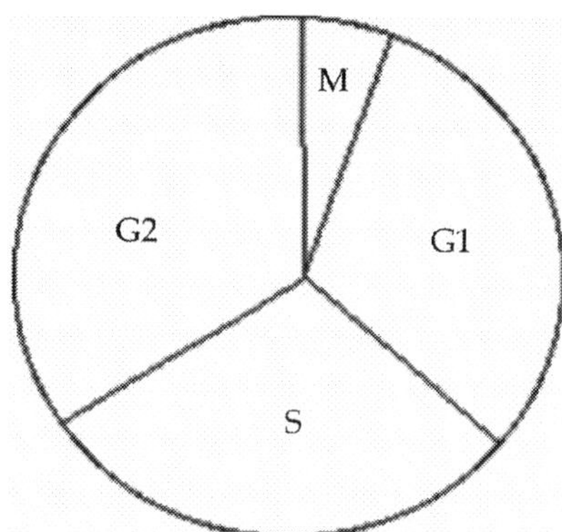

Cells which are not dividing are said to be in the G0 phase

GENETIC BLUEPRINTS FOR CELLS ARE ORGANIZED INTO CHROMOSOMES

The plans for making cells are coded in DNA

DNA is organized into giant molecules called chromosomes

Each chromosome is a single DNA molecule containing many genes

Each gene gives the directions for making 1 protein

In humans each chromosome has approximately 2000 genes

Chromosomes have distinct parts

Centromeres:

Hold duplicated chromosomes together before they are separated in mitosis

Kinetochore proteins bind to centromere and attach chromosome to spindle in mitosis

Telomeres: ends of chromosomes: important in cell aging

DNA in chromosomes is associated with proteins

Proteins strengthen DNA fiber

Package chromosomes when they condense

Control activity of genes

Each Species Has a Set Number of Chromosomes

Humans have 46 chromosomes (23 pairs)

One chromosome from each pair comes from the father, the other comes from the mother

46 is the diploid number, present in body (somatic cells)

Germ cells (sperms and eggs) have 23 chromosomes (they are not paired)

Each species has a specific chromosome number (these are the diploid numbers):

Most bacteria have 1 circular chromosome

Fruit fly: 8

Dog: 78

Amoeba: about 250

Pea plant: 14

Comparison of the Types of Cell Division

Terms for the number of pairs of chromosomes:

Haploid: 1 pair

Diploid: 2 pairs

Polyploid: many pairs

Terms for the relationships between chromosomes:

Homologous chromosomes: 2 chromosomes of the same type, one from the mother and one from the father

Sister chromatids: 2 identical chromosomes, formed by duplication, still held together by their centromeres (see drawing on next page)

Binary fission:

Prokaryotic cells (bacteria)

DNA duplicates t 2 identical chromosomes

In the figures below 1D, 2D and 4D refer to the relative amounts of DNA

Each chromosome attaches to the plasma membrane

As cell grows chromosomes are pulled apart

Makes 2 identical cells

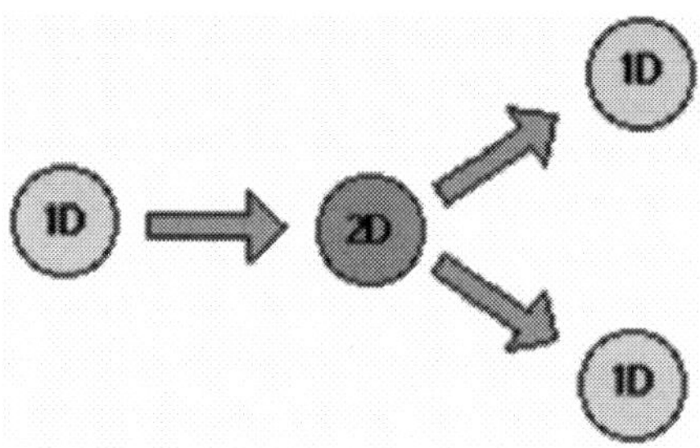

Mitosis:

Eukaryotic cells

DNA duplicates → 2 sister chromatids

Chromosomes attach to spindle and separate

Used for growth, repair and reproduction (in single-cell organisms)

Makes 2 identical cells

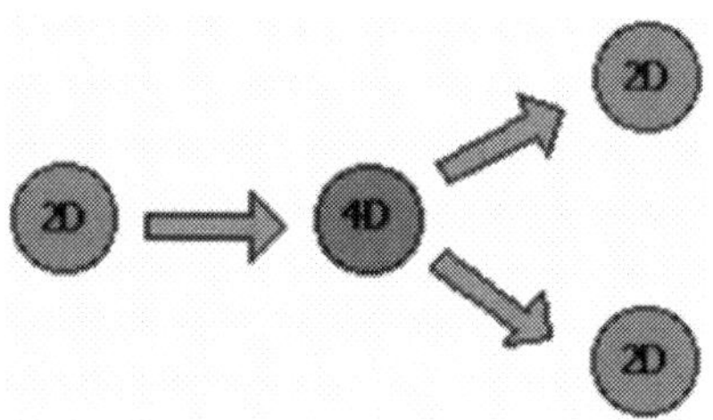

Meiosis:

Eukaryotic cells

DNA duplicates → 2 sister chromatids

Chromosomes attach to spindle & separate

Two divisions

First separates homologues

Second separates sister chromatids

Used for sexual reproduction (makes sperm & eggs)

Makes 4 haploid cells (half the number of chromosomes)

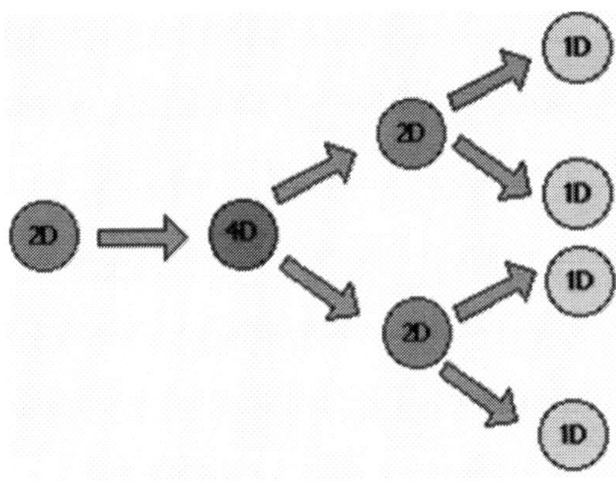

Interphase: Before a Cell Can Divide it Must Duplicate its Chromosomes

To make a new cell the old cell must duplicate all its parts

Duplication takes place in interphase

DNA (chromosomes) duplicated in the S subphase

The duplicated chromosome remains attached to the original chromosome by its centromere

The original chromosome and its duplicated partner are called sister chromatids

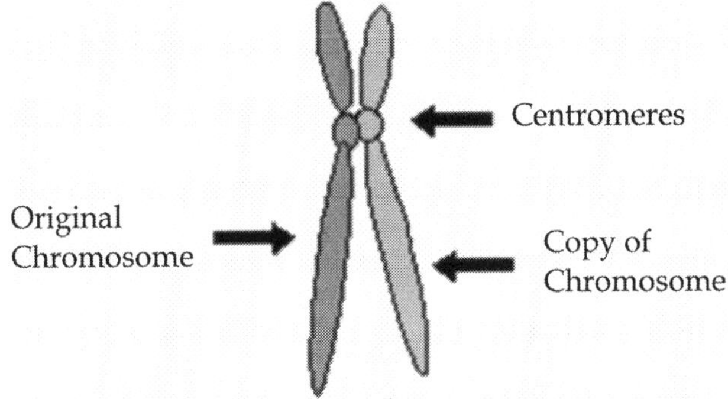

Other parts (proteins, organelles) duplicated in G1, G2 and S

To duplicate the various parts DNA must be in an unwound state

DNA very thin when unwound, cannot be seen in light microscope

Most of the cell cycle (~98%) is interphase

The rest of the cell cycle (~2%) is called mitosis

Prophase

Prometaphase

Metaphase

Anaphase

Telophase (& Cytokinesis)

Prophase: the Chromosomes Must Condense Before they Can be Separated

Interphase chromosomes are very long and thin

Very fragile

Would tangle if you tried to separate them in this state

Must be packaged before they can be separated

In packaging chromosome combines with proteins

Length is reduced by 10,000 times

Chromosomes become thicker and this makes them visible under the light microscope

Excellent diagram of packaging on p. 353 of text

Packaging takes place at the beginning of mitosis: prophase

In prophase the mitotic spindle begins to form

The Mitotic Spindle is Finished in Prometaphase

Chromosomes separate on a protein scaffold called the mitotic spindle

Made of tubulin protein → microtubules

Centrosome (don't mix this up with the centromere) organizes construction of spindle

In animal cells the centrosome has a "9 + 2" structure, the centriole, in the center

Function of centriole unknown: apparently not required for cell division

Centrosome divides and the 2 copies move to opposite sides of the cell

Tubulin fibers radiate out from centrosomes, meet in center of cell

Chromosomes attach to some of the spindle fibers by kinetochores:

Attachment: body of chromosome → centrosome → kinetochore -> spindle

Nuclear membrane breaks down at beginning of prometaphase

Metaphase: the Chromosomes Line Up in the Center of the Cell

To separate the chromosomes the 2 sister chromatids must attach to fibers going to opposite ends of the cell

In metaphase the chromosomes line up in the center of the cell, but the sister chromatids remain attached to each other

Anaphase: the Chromosomes Must be Exactly Split Between 2 Cells

In anaphase the centromeres split apart, releasing the sister chromatids

The chromosomes move rapidly toward the poles (centrosomes) of the cell

Movement mainly caused by shortening of spindle fibers connected to kinetochores

Cytokinesis & Telophase: the Cell Splits into 2 New Cells

Once the chromosomes are separated into 2 groups the nuclear membrane reforms

In animal cells the cell is pinched into 2 cells by a contractile band in the center (cytokinesis)

Produces a cleavage furrow

Plant cells are too stiff to be pinched apart- cell plate formed instead, separates old cell into 2 new cells

The Cell Division Cycle is Regulated by Cyclin Proteins

Cell cycle is highly regulated, with "checkpoints"

Cell must be a certain size before it can enter mitosis

Mitosis will not start unless the DNA is completely duplicated

Timing of the cycle is controlled by a group of proteins called cyclins

Cyclins bind to and activate protein kinase enzymes (Cdks); the activated complex is called MPF

Protein kinases put phosphate groups on proteins, which switches their activity on or off

Negative feedback causes the cyclins to break down; this causes the system to oscillate, producing the cell cycle

Summary of Cell Cycle

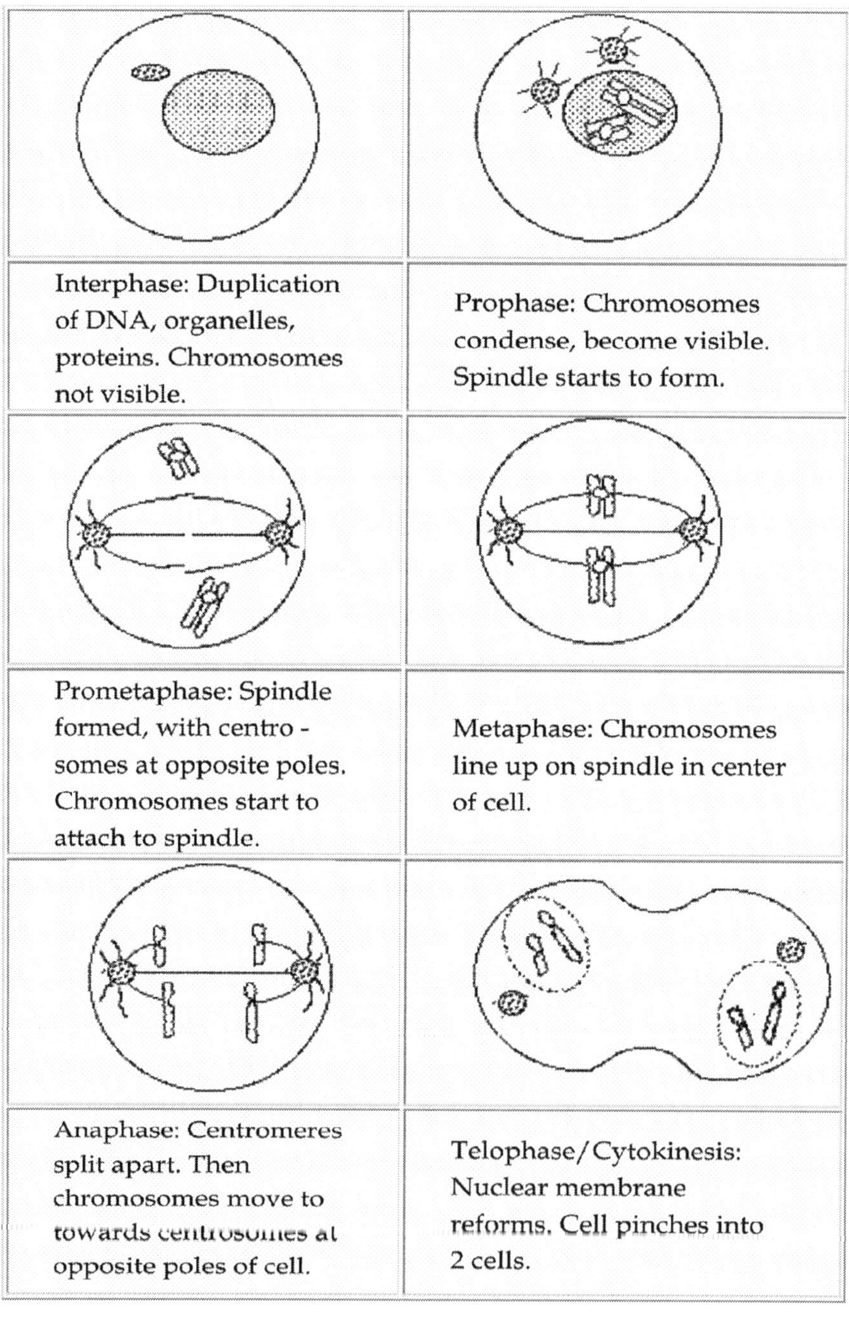

CHAPTER - 13

Meiosis : Sexual Reproduction

SEX IS VERY COSTLY

Large amounts of energy required to find a mate and do the mating: specialized structures and behavior required

Intimate contact provides route for infection by parasites (AIDS, syphillis, etc.)

Genetic costs: in sex you pass on only half your genes to your children

Males are an expensive luxury- in most species they contribute little to rearing offspring

But There are Some Advantages

More genetic diversity: more potential for survival of species when environmental conditions change

Shuffling of genes in meiosis (random separation of homologous chromosomes)

Crossing-over in meiosis (see below)

Fertilization (genes from 2 individuals brought together)

DNA back-up and repair

Asexual organisms don't have back-up copies of genes, sexual organisms have 2 sets of chromosomes and one can act as a back-up if the other is damaged

Sexual mechanisms , especially recombination, are used to repair damaged DNA- the undamged chromosome acts as a template and eventually both chromosomes end up with the correct gene

Reproduction Without Sex and Sex Without Reproduction Both Occur in Nature

Sex is the transfer of genes from one cell to another and in microorganisms this often occurs without cell division, so that there is no reproduction

Bacteria can transfer genetic material through projections called pili

Many species can reproduce without sex

Most single-cell organisms (bacteria, protista)

Some multicellular organisms (i.e., whiptail lizards, hydra, sea anenomes, aphids)

Terminology of Meiosis

Review of diploid & haploid

Haploid = single set of chromosomes

Diploid = double set of chromosomes (one from each parent)

Homologous chromosomes: Pairs of chromosomes having the same genes. Diploid cells have homologous chromosomes (one from each parent).

Sister chromatids: a chromosome and its identical copy (formed by DNA duplication), still held together by centromeres.

In Meiosis 2 Cell Divisions Cut the Number of Chromosomes in Half

Meiosis is designed for sexual reproduction

Parents have 2 sets of chromosomes in their somatic cells; they are diploid

Sperms and eggs must have only 1 set; they must be haploid

When a sperm and egg combine in fertilization they create a new individual with 2 sets of chromosomes (diploid)

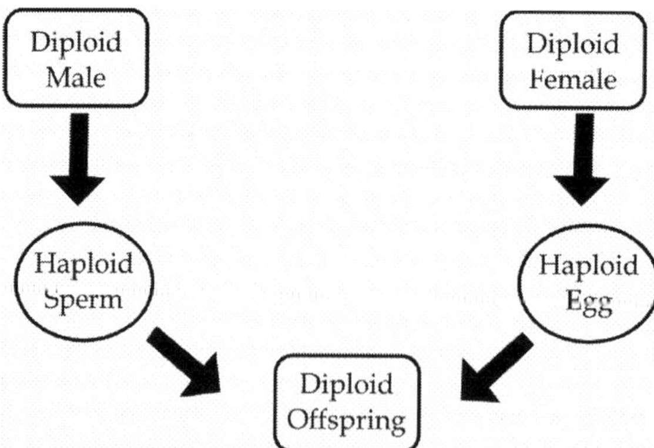

Meiosis creates the sperm and egg cells by reducing the number of chromosomes to half (1 set)

Starts with DNA duplication in 1st interphase

No DNA duplication in 2nd interphase

Review diagram of cell DNA in meiosis

The First Cell Division Separates Homologous Chromosome

In 1st prophase and prometaphase the homologous chromosomes come together (**synapsis**) and attach to the spindle together through a protein complex

The association of the 2 homologous chromosomes is sometimes called a tetrad

In the first meiotic division the centromeres remain attached and the sister chromatids stay together

The attachment between the homologous chromosomes breaks down and the 2 homologues are pulled apart

This gives 2 cells, each with 1 set of duplicated chromosomes (the cells are considered haploid)

The stages of cell division are the same as those in mitosis (prophase, prometaphase, metaphase, anaphase, telophase) except that the prophase/ prometaphase period is more complex due to the association of the homologues

At the end of the first meiotic division the cell does not go back into a long interphase

There is no DNA duplication before the 2nd meiotic division

In some species the nuclear membrane reforms between divisions; in other species it does not

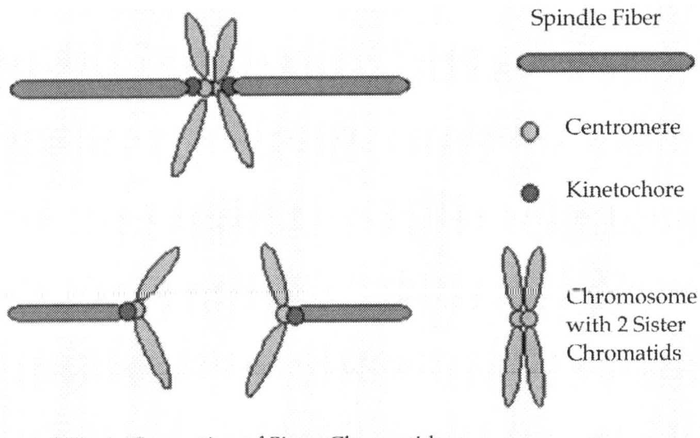

Mitosis: Separation of Sister Chromatids

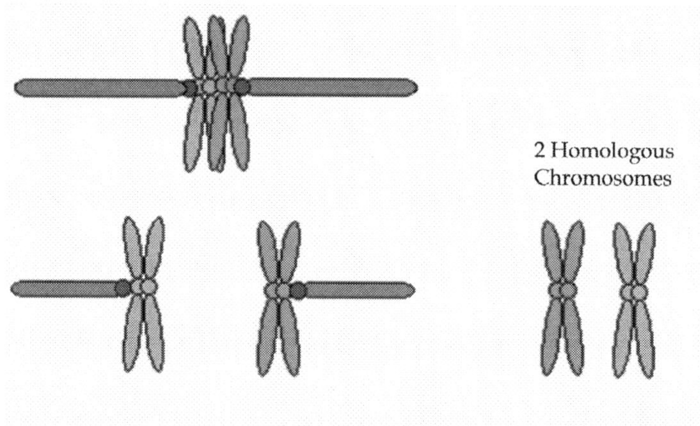

1st Meiotic Division: Separation of Homologues

The Second Cell Division Separates Sister Chromatids

The 2nd meiotic division is like mitosis

The chromosomes line up at metaphase with sister chromatids attached to spindle fibers going in the opposite directions

At anaphase the centromere attachments break and the sister chromatids are pulled to opposite sides of the cell

At the end of 2nd meiotic division there will be 4 haploid cells (sperms or eggs)

Synapsis and Crossing-Over Occur in Meiosis

In the 1st meiotic division the homologous chromosomes attach to one another by a protein complex

This brings chromatids from the 2 homologues into contact with each other and sometimes they cross

Cross-overs help to hold homologous chromosomes together

At the cross-over the 2 chromatids can exchange tips (recombination)

The exchange is catalyzed by enzymes

Promotes genetic diversity

Homologous chromosomes can have more than 1 cross-over

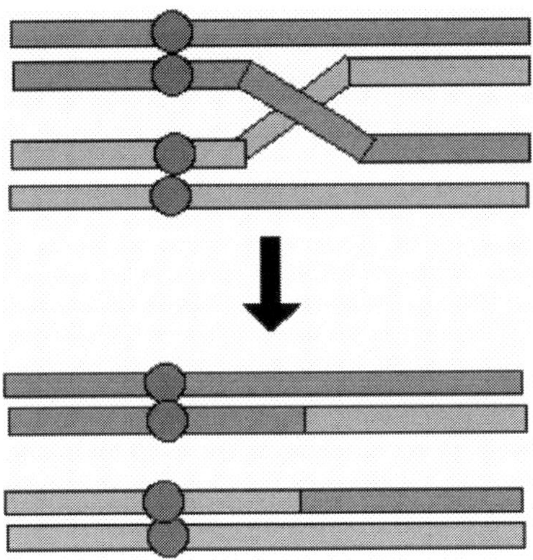

The cross-over process is unbelievably precise

Average human chromosome has about 100 million DNA base pairs

In cross-overs neither chromosome can gain or lose a single base pair

Meiosis Can be Interupted for Long Periods of Time in Females

When meiosis is used to produce egg cells the process is often interupted for long periods (up to a year or more)

Vertebrate females produce all of their potential egg cells (primary oocytes) before birth

These cells have started meiosis, but the process has been arrested in the first prophase of meiosis

The homologues are paired, with crossing-over of their chromatids

In humans the arrested state lasts for about 15 to 50 years

Chromosomes partly unwind (decondense) in the arrested state and the potential egg cell matures

Later in development, under the influence of hormones, some of the primary oocytes finish the first cell division and start the second

First cell division is very uneven: the smaller of the 2 cells produced becomes a "polar body" and dies

The oocyte (now called a secondary oocyte) arrests division a second time, usually at the 2nd metaphase of meiosis

At ovulation the secondary oocyte is released from the ovary

If it is fertilized it finishes the 2nd meiotic division; otherwise it dies

Again a small polar body is produced and is discarded

The remaining egg nucleus fuses with the sperm nucleus and development of a new individual begins

Some Species Alternate Haploid and Diploid Generations

Animals are primarily diploid, with haploid sperms and eggs

Many plants have long-lasting haploid structures, alternating with diploid structures (alternation of generations)

Examples:

In mosses the haploid plant is the dominant structure that you see; the diploid sporophytes can be seen rising on stalks above the main plant

In ferns, conifers and flowering plants the most conspicuous structure is diploid; the fern has a tiny haploid gametophyte and in conifers and flowering plants only the pollen, egg and a few cells surrounding the egg are haploid

Genetic Defects Can be Caused by Accidents in Chromosome Separation

Sometimes mistakes are made in meiosis and one cell gets an extra chromosome and the partner cell gets one less

Comparison of Mitosis and Meiosis

Things common to both meiosis & mitosis:

Both start with DNA duplication in interphase

Use of a mitotic spindle to separate the chromosomes

Division with a set of stages: prophase, prometaphase, metaphase, anaphase, telophase

Unique events and results of mitosis:

Produces 2 identical cells

Does not promote genetic variability

Used for growth and repair (reproduction in single cell organisms)

A single cell division

Separates sister chromatids

Centromeres divide at anaphase

Unique events & results of meiosis:

- Produces 4 haploid cells (with half the original number of chromosomes)
- Promotes genetic diversity
- Used for sexual reproduction (produces sperms and eggs)

Two cell divisions

- First division separates homologous chromosomes
- Second division separates sister chromatids
- Centromeres divide at 2nd anaphase but not at 1st

No duplication of DNA between 1st and 2nd divisions

Homologous chromosomes come together (synapsis) in 1st division

Crossing-over between homologous chromosomes common

CHAPTER - 14

Mendelian Genetics

A fair number of practical breeding experiments were done in ancient times

People knew there was a connection between sex and reproduction

The idea had developed that "like begets like"

By selecting certain plants and animals and breeding them many domestic varieties of plants and animals had been developed

Example: corn was developed from a wild grass in Mexico and Central America, probably about 5000 years ago

149 of the top 150 plant crops were domesticated by primitive man

Another example: dogs were domesticated by 12,000 years ago; selective crossing has given us all the many breeds we have today

The role of pollination in producing fruit had been understood for several thousand years and some artificial pollination was done to increase yields

Sperm were discovered by Anton van Leeuwenhoek around 1680

Eggs were not observed in mammals until 1827

Chromosomes had been seen in the 1800s, but their role in heredity was not understood until the early 1900s

THE IDEA OF DISCRETE HEREDITARY ELEMENTS (GENES) WAS DEVELOPED BY AN AUSTRIAN MONK, GREGOR MENDEL

Gregor Mendel (1822-1884) was a monk at a monastery in Brunn, Austria

He taught school and did other chores, but in his spare time he set out to study plant breeding

He started breading pea plants about 1855

He wanted to learn the rules of heredity so that better varieties could be produced

Mendel Artificially Pollinated Pea Plants and Studied the Results

To breed the peas Mendel atificially pollinated them

Like most flowers, pea flowers are complete, with both male and female parts

Flower parts:

Carpel (female):

Pistil: "landing platform" for pollen

Style: pollen tube grows from pistil to ovary

Ovary: makes egg cells

Stamen (male):

Anther: makes pollen (2 sperm + accessory cells)

Filament: attaches anther to base of flower

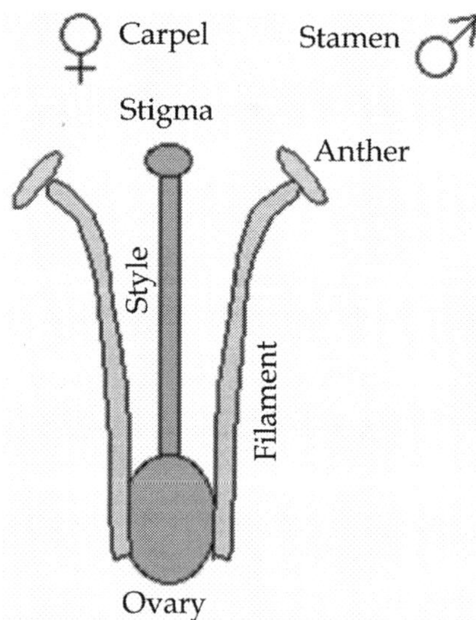

To prevent self-fertilization of the pea flowers Mendel pulled off the stamens with forceps

He artificially fertilized the flowers by putting pollen from another flower on the stigma with a brush

Covered flowers with bags to prevent fertilization by pollen crried by bees or wind

Collected the seeds produced by the flowers and planted them to see the results of his crosses

Mendel Picked 7 Characters to Study: Each Had 2 Variations (Traits)

By repeatedly crossing pea flowers and selecting for certain traits Mendel developed pure lines for these traits

Example: flower color:

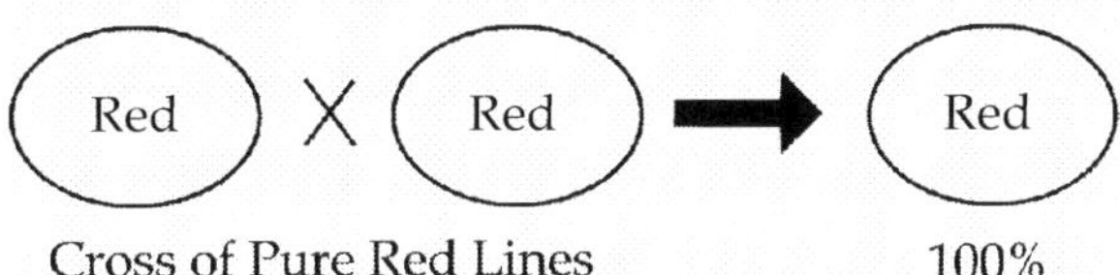

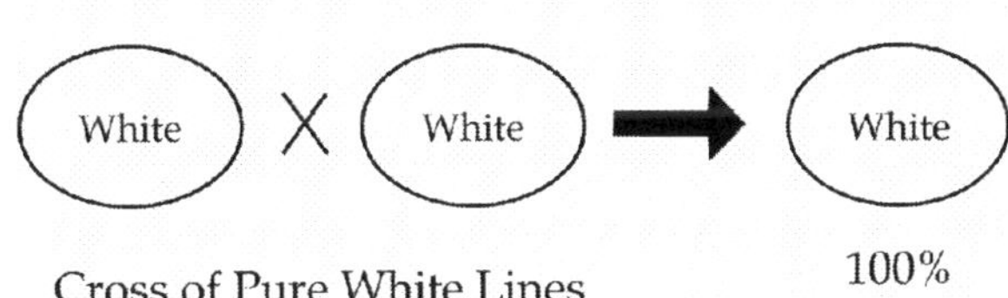

He decided to study 7 flower characters, each of which had exactly 2 traits

Character	Traits
Flower color	Purple
	White
Flower position	Axial
	Terminal
Seed color	Yellow
	Green
Seed shape	Smooth
	Wrinkled
Pod color	Green
	Yellow
Pod shape	Inflated
	Constricted
Stem length	Tall
	Short

In modern terms the traits are callled **phenotypes** (what you see)

The genes producing the phenotype is called the **genotype**

When One Pure Line Was Crossed with Another Pure Line

One trait was dominant

We will consider Mendel's crosses of flowers of different color

When Mendel crossed pure reds with pure whites the surprising result was that the offspring were 100% red

Note on terminology:

The original pure lines are called the parent generation

The offspring (hybrids) are called the F1 (first filial) generation

Most people would have expected a blended color such as pink, but the red color was clearly dominant

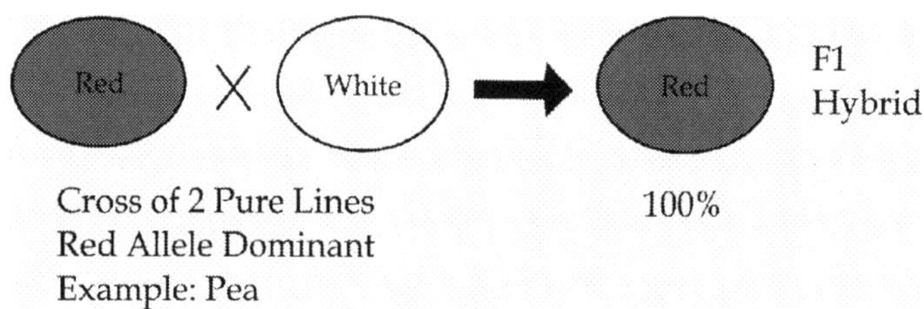

Cross of 2 Pure Lines
Red Allele Dominant
Example: Pea

Had the cross destroyed the hereditary factor producing white?

To get more information Mendel decided to cross the hybrids (this gives an F2 generation)

When he did this some white flowers reappeared!:

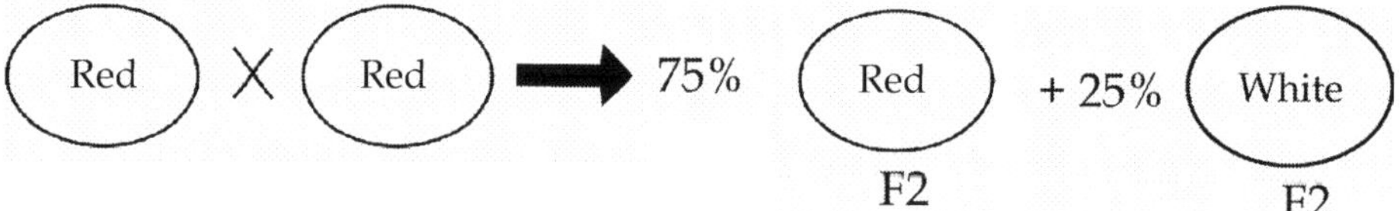

Cross of F1 Hybrids
Recessives Reappear
in F2 Generation

The white element had not been destroyed by the first cross, it was just not seen because the red element was dominant

Mendel proposed that there were 2 hereditary elements, one producing red and the other producing white

In modern terminology the flower color is caused by a **gene** with 2 **alleles,** one for red and the other for white

The red element is dominant over the white element

The ratio of the traits (phenotypes) in the F2 generation is always 3 dominants to 1 recessive

Note on notation:

One notation for genes uses a capital letter for a dominant gene, such as R for red

The recessive gene is given the same letter, but in lower case, such as r for white

This is called a monohybrid cross (1 character, flower color)

Note on tampering with history:

Mendel's flowers were purple and white, but I have changed them to red and white

Why? Because it is much easier to tell the difference between R & r, than between P & p, especially in handwritten notes

There are 2 types of red flowers, pure and hybrid:

One has 2 red genes, RR

One has 1 red and 1 white gene, Rr

There is only 1 type of white flower

White flower cannot have R gene, must be rr

Phenotype(trait)	Genotype(kind)	
Red	RR	Pure: homozygousdominant
	Rr	Hybrid: heterozygous
White	rr	Pure: homozygousrecessive

All of the other traits studied by Mendel followed the same pattern

When the pure lines were crossed one trait was 100% dominant over the other

When the F1 hybrids were crossed the F2 generation had a dominant/ recessive ratio of 3/1

Mendel Proposed that Genetic Elements (Genes) Segregate in the Formation of Pollen and Ovules

To explain his results Mendel proposed that the genetic elements were segregated (separated) when eggs and sperm were formed, so that each egg or sperm received only 1 or the 2 elements

The elements are now called genes

Variations of the same gene are called alleles

A diploid cell can have only 2 alleles (one on each of the homologous chromosomes)

Half the gametes (sperms & eggs) get one allele, half get the other

In fertilization the new zygotes would get 2 of the elements (one from the sperm, one from the egg) in a random distribution

This can be shown in Punnet diagrams:

The left circles show the types of sperms that are possible

The right circles show the types of eggs that are possible

The squares show the results of crosses

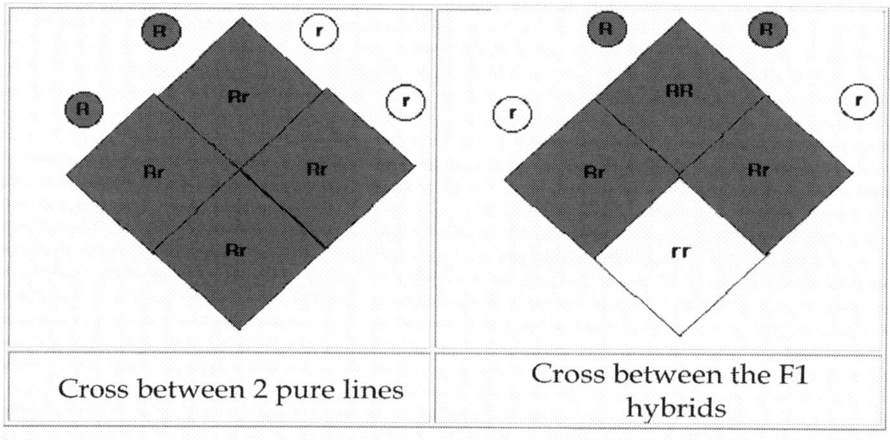

Cross between 2 pure lines	Cross between the F1 hybrids

As you can see, segregation of the alleles explains Mendel's experiments nicely

We now know that segregation occurs because the alleles are on homologous chromosomes and the homologues are separated in meiosis

The Genotype Can be Determined with Testcrosses

Suppose you have a red flower but you don't know if it is a pure red (RR) or a hybrid (Rr)

You can tell which one by crossing it with another flower

What flower would you use for your testcross?

What result would you get if the unknown flower is pure red?

What result would you get if the pure flower is hybrid red?

The Dihybrid Cross Illustrates the Principle of Independent Assortment of Alleles

After working out the principle of segregation of the alleles Mendel did crosses in which he studied 2 characters at the same time (dihybrid crosses)

As an example consider the 2 characters flower color and height:

Character	Dominant Phenotype	Recessive Phenotype	Dominant Allele	Recessive Allele
Height	Tall	Short	T	t
Flower color	Red	White	R	r

Mendel crossed pure tall red plants (TTRR) with pure short white plants (ttrr) to get 100% F1 hybrids (TtRr) that were tall red

He then crossed the hybrids with themselves

He got 4 phenotypes: tall red, tall white, short red and short white

The ratio of the phenotypes was 9:3:3:1, as shown in the table

Color Code	Phenotypes	Genotypes	Ratio
	Tall red	TTRR, TTRr, TtRR, TtRr	9
	Tall white	TTrr, Ttrr	3
	Short red	ttRR, ttRr	3
	Short white	ttrr	1

He explained the ratio by assuming that the genetic elements (alleles) for different characters were sorted independently of each other when the eggs and sperm are formed

Remember that each sperm or egg must have a gene for both height and flower color.

If the genes for height and flower color sort independently there are 4 possible sperms or eggs: TR, Tr, tR and tr

If you cross these eggs and sperms you will get a 9:3:3:1 ratio of phenotypes in the F2 generation, as shown in the Punnet diagram below

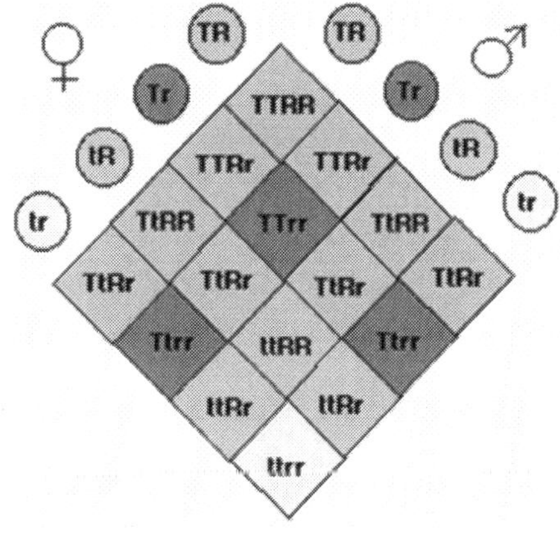

To see this use the color code from the table above and count the diamonds

The modern explanation for this is that the genes sort independently because they are on different chromosomes

As we will see, if genes are on the same chromosome they will not sort independently

Mendel Was a Good Experimental Scientist but Also Had Some Luck

Mendel succeeded for a number of reasons:

He had excellent experimental technique

He kept detailed records of his crosses for several years

He collected numerical data- it was the ratios of the crosses that clinched the arguments for his theories

He was good at mathematical analysis

He picked a simple system with a few well defined characters to study

His characters all showed dominant/recessive traits- this made his analysis much easier

He was lucky- some of his characters were on the same chromosome, but were so far apart that crossing-over made them sort nearly independently

Mendel's Terminology	Modern Terminology
Hereditary elements	Genes with alleles
Trait	Phenotype
Kind	Genotype
Hybrid (Aa)	Heterozygous
Pure (AA or aa)	Homozygous

CHAPTER - 15

Genetics : Beyond Mendel

MENDEL'S RULES OF GENETICS DO NOT ALWAYS WORK

Mendel's rules:

Segregation of alleles

Independent assortment of genes for different characters

Hold for characters:

Which have dominant & recessive traits

Determined by single genes

Genes are on separate chromosomes

Different types of chromosomes, or

Different homologues of same type of chromosome

These conditions are not always met, so there are many complications

Alleles Are Not Always Dominant & Recessive

All of the characters studied by Mendel had traits that were dominant or recessive

For many characters the 2 alleles give a blended result

Example: flower color in snapdragons

Blending of this sort is called incomplete dominance

In another type of blending both alleles seem to assert their whole effect; example- blood group antigens (see multiple alleles section below); this is called codominance

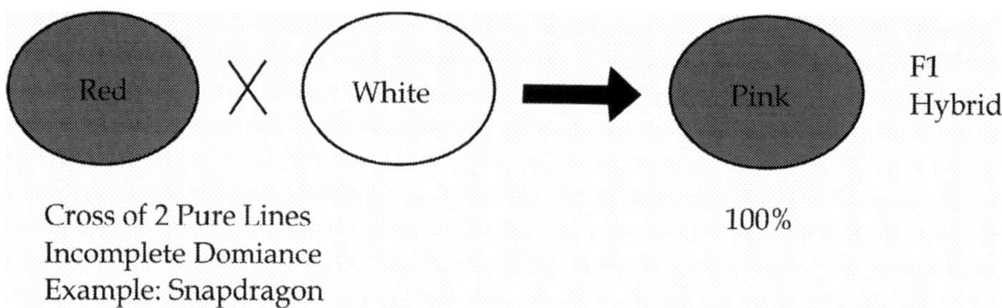

Dominance may depend upon which function of the gene is monitored: Example sickle cell anemia gene

Sickle cell gene acts like a recessive in producing anemia

Hemoglobin mutation, causes hemoglobin to form crystals, damaging red blood cells

AA = normal (A is normal hemoglobin allele)

AS = heterozygous, relatively normal (no anemia)

SS = homozygous for sickle gene, causes severe anemia

If you look at the amount of hemoglobin produced the alleles look like they are codominant

AS person produces 50% hemoglobin A, 50% hemoglobin S

A Character May Be Determined by More Than One Gene

Probably most characters are the result of many genes (i.e., skin color or height in humans)

A good example is the feather color of budgies (shown in text, p. 246)

Color caused by 2 genes: Y gene produces yellow pigment in outer feather; B gene produces bluish pigment (melanin) in core of feather

YYbb or Yybb → yellow bird

yyBB or yyBb → blue bird

YYBB, YyBB, YYBb or YyBb → green bird

yybb -> white bird

Example 2: coat color of Labrador retrievers (a similar case is discussed in the text, p. 250)

Color involves 2 genes, B & E

B produces color; E deposits color in coat

BB or Bb → black color; bb → chocalate brown color

EE or Ee → color deposited in coat, dogs have color determined by B gene; ee → color not deposited in coat, dogs are yellow

Do dihybrid crosses for the budgie and labrador retriever

Budgie: cross F1 hybrids (YyBb X YyBb)

What are the phenotypes and what are the ratios?

Labrador: cross F1 hybrids (BbEe X BbEe)

Again, what are the phenotypes and what are the ratios?

Does one of the ratios surprise you

There May be More Than 2 Alleles in a Population

An individual can have only 2 alleles of a gene, one on each homologous chromosome

In a population there can be more than 2 alleles

Example: ABO blood groups in humans

Important because ABO blood groups affect blood transfusions

Genes cause expression of sugar groups on surface of red blood cell membrane; these carbohydrates act as antigens in immune reactions

3 alleles: A, B, O

A & B dominant over O

A & B codominant to each other

Type O produces no sugar antigens

6 genotypes and 4 phenotypes

Phenotype	Genotypes
A	AA or AO
B	BB or BO
AB	AB
O	OO

Phenotype can be determined by treating drops of blood with antisera which react with the A or B sugar groups

If the sugar groups are present the antisera will cause the red cells to clump together

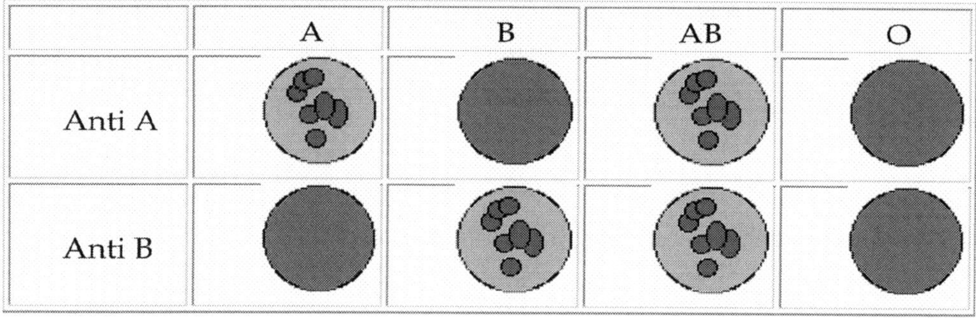

You have blood serum antibodies directed against blood groups not found on your red blood cells

Blood Type	Serum Antibodies
A	Anti B
B	Anti A
AB	None
O	Anti A & Anti B

The ABO antigens are important in blood transfusions; transfusion of the wrong type of blood will cause red cells to clump, can be lethal

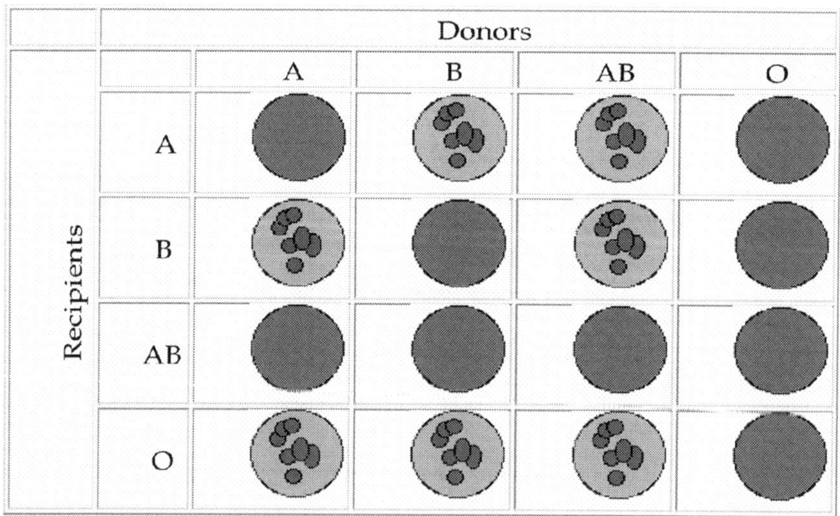

To determine transfusion reaction think of the recipient serum attacking the blood cells of the transfusion

Why is type O considered the universal donor? What type is considered the universal recipient?

Different cultural groups have different percentages of blood groups

Environmental Conditions Can Control the Expression of Heredity

The environment sometimes affects the expression of a gene, changing phenotypes

Example: Himalayan rabbit

Coat color in a warm environment (above 30 deg C) is white

At 25 deg C colder parts of the animal (ears, tip of nose, feet) are black

Melanin-producing gene is turned on in cells at low temperature

Different Genetic and Non-Genetic Systems Are Used to Determine Sex

In mammals (including humans) sex is determined by the 23rd chromosome pair

Not a true pair- a large X chromosome and a small (probably degenerate) Y chromosome

Partially homologous, pair up in meiosis

XX = female; XY = male

Y has a gene necessary for making testosterone

In birds individuals with the same sex chromosome are males (ZZ); those with different sex chromosomes are females ZW

In bees and ants unfertilized eggs become males (haploid), while fertilized eggs become females

In the fruit fly XX gives a female and XY becomes a male, but the situation is more complicated than in mammals; it is really the ratio of X chromosomes to somatic chromosomes which determines sex

In turtles temperature determines whether an egg hatches as a male or a female

In some species, such as aphids and certain coral reef fishes, females can turn into males if the need arises

In a few species one of the sexes has been lost:

Whiptail lizards are 100% female; diploid egg develops without sperm

Some salamanders, *Daphnia*, rotifers, bees, ants also parthenogenic

Nematode, *Caenorhabditis elegans*, has no true females, only males and hermaphrodites (XX = hermaphrodite; XO = male)

Chapter - 16

Chromosomes and Inheritance

Chromosomes are giant molecules of DNA
Found in nucleus: never leave except in cell division
Carry hereditary units, genes
Two homologous copies in diplod cells
Unwind when replicating or making RNA
Wind up (condense) in cell division
Attach to spindle and separate in meiosis & mitosis
Crossing over in prophase of 1st meiotic division
Chromosomes are equally distributed in mitosis, meiosis
Gametes have half the number of chromosomes as body cells
Sister chromatids are observed to split apart longitudinally in anaphase

If Genes Are on the Same Chromosome the Genes Are Linked

In the human genome there are an estimated 50,000 to 100,000 genes, but only 23 types of chromosomes
Each chromosome must have many genes (about 2000 to 4000 average)
Genes on the same chromosome are physically linked to each other

Genes Which Are Linked Will Not Sort Independently

If genes are lined they will sort together, not independently in meiosis
This reduces the number of phenotypes and changes the genetic ratios
Compare genes for 2 characters, A & B

Let A & B be dominant to a & b

In the unlinked genes B & are on separate homologous chromosomes; A & a are on another set of homologous chromosomes

In the linked situation let the A and b genes be on one chromosome, while the a & B are on the homologue

If the genes are on different chromosomes they will assort independently in meiosis and make 4 types of gametes

The linked genes cannot separate in meiosis, there will be only 2 types of gametes; gametes with Ab or aB together are impossible (except with crossing-over- see below)

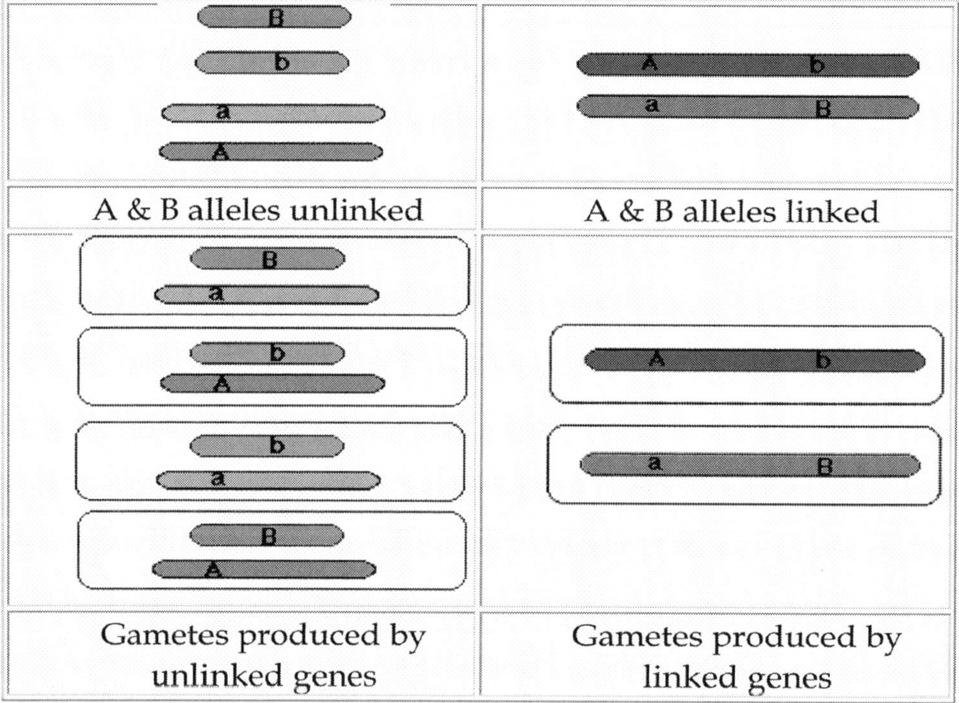

Unlinked genes give a wider range of genotypes and phenotypes

To see this consider crossing AaBb hybrids which are linked and unlinked

The cross will be AaBb X AaBb

The gametes will be those shown in the table above

Crossing-Over Increases Genetic Diversity

In the 1st meiotic prophase homologous chromosomes are brought together and their strands can exchange genes by crossing-over

In this way linked genes become unlinked to some extent

Example: Tomato plants (this example is taken from The *Cartoon Guide to Genetics* by Larry Gonick & Mark Wheelis)

Have gene controlling height: Tall (T) dominant over short (t)

Also have a skin texture gene: Smooth (S) dominant over hairy (s)

Suppose a hybrid tall smooth tomato (TtSs) is crosssed with a pure recessive (ttss); this is a testcross

If the T and S genes are linked with both recessives on one homologous chromosome (TS) and both recessives on the other (ts), the expected genetic ratios will be different from those expected for unlinked genes

The results of the cross should be a test for linklage

Phenotype	Expected %if Linked	Expected %if Unlinked	Actual Results
Tall smooth	50	25	48
Tall hairy	0	25	2
Short smooth	0	25	2
Short hairt	50	25	48

The actual results are close to the expected percentages if the genes are linked; the 4% difference between observations and expectations is due to crossing over

If 2 genes are far apart they will cross over frequently; if they are close together cross-overs will be rare

This result can be used to map genes

Mistakes in Meiosis Can Produce Gametes with Extra Chromosomes

Sometimes chromosomes do not separate properly in meiosis

Called nondisjunction

This produces one gamete with an extra chromosome and one gamete with a reduced number

Almost always this is lethal

Exceptions to lethality:

Extra chromosome 21 → Down syndrome (various defects, including mental retardation)

Extra X chromosome

XXX: normal female

XXY: Klinefelter's syndrome- male, usually sterile, normal intelligence

Single X chromosome: XO: Turner syndrome, female, usually sterile, usually normal intelligence (YO is lethal; at least one X is required for life)

X-Linked Genes Complicate Mendel's Rules

Linkage was first observed on sex chromosomes

Usually sex-linked genes are located on the X chromosomes- Y chromosome is degenerate and does not have all of the genes found on the X

Females, with 2 X chromosomes, will tend to follow Mendel's dominance/ recessive rules (but they are complicated by X chromosome inactivation, see below)

Males have a single X

Recessives on the X chromosome will be expressed in the male more ofthen than in the female; male will not have a dominant gene to oppose the recessive

If 1/1000 of X chromosomes have a recessive gene the recessive phenotype will appear in 1 out of a thousand males (since they have only 1 X) but in only 1 out of a million females (since they have 2 Xs)

Some X-linked recessive diseases:

Duchenne muscular dystrophy

Hemophilia A

Red/green color blindness

Heredity of X-linked genes:

Males get their X chromosome from their mothers (they get a Y from their fathers); X-linked recessive genes must come from the mother

Males donate any X-linked recessives they have to their daughters, who then become "carriers" of the recessive

Pattern: mother → son → daughter

In the figures below **Xr** denotes an X chromosome with a recessive gene.

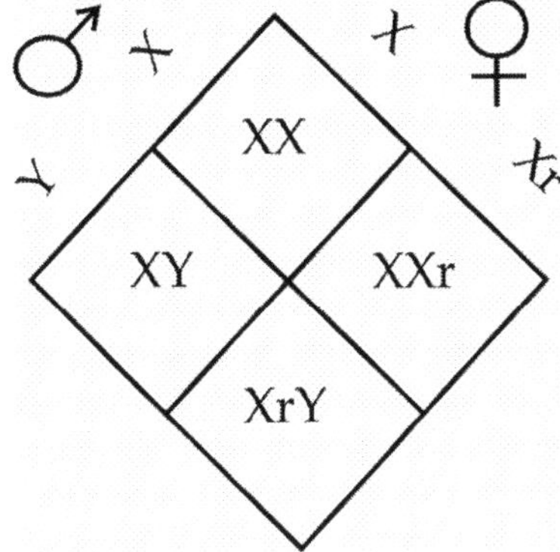

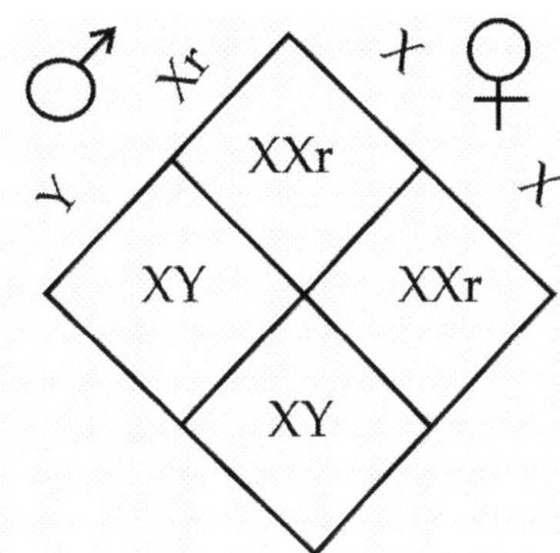

Female carrier and normal male: female passes gene to half of her sons and daughters. Only the son shows the phenotype, however. The daughter has a dominant gene to block the recessive. She is a carrier.

Normal female and affected male: the male passes his recessive to all of his daughters, who become carriers. His sons are normal because they get their X chromosomes from their mother.

Genetic Determination of Sex Creates Dosage Problems

In mammals females have 2 X chromosomes, the males only 1

If nothing were done to compensate the females would get a double dose of any gene products from the X chromosome, compared to the dose that males get

Nature solves this problem by shutting down one whole X chromosome in mammalian females

X chromosome inactivation is called Lyonization after Mary Lyon who discovered it

Inactive X chromosome appears in condensed state as a Barr body

Inactivation of X chromosomes in different cells is somewhat random

The calico cat is a product of X chromosome inactivation

Genes for coat color of the cat are on the X chromosome

One gene produces a black color; its allele produces orange

To get a calico coat a cat must be heterozygous, with genes for both the orange and the black color

If the X chromosome with the black gene is inactivated that cell will produce orange

If the X chromosome with the orange gene is inactivated the cell will produce black

Inactivation occurs in patches, giving the orange and black coat of the calico

CHAPTER - 17

DNA and Inheritance

UNTIL THE 1940S BIOLOGISTS ARGUED ABOUT WHETHER DNA OR PROTEIN WAS THE MOLECULE OF HEREDITY

Nuclei observed to divide when cells reproduced themselves

Large amounts of both DNA & protein in the nucleus

Many though protein was most likely molecule of heredity

Proteins made of 20 different amino acids- could have more variety than DNA (made of only 4 different bases)

Pneumonia Bacteria Have a Transforming Factor

Pneumonia bacteria (*Streptococcus pneumoniae*) exist in 2 forms:

R (rough): harmless

S (smooth): "wild type", causes disease

The S type is coated with a polysaccharide which makes it infective and gives the colonies a smooth appearance

Frederick Griffiths about 1928 studied the R & S strains by injecting them into mice

S injected into mice → pneumonia → death

R injected into mice → harmless

Also, boiled S injected into mice → harmless (bacteria killed by boiling)

The Griffiths did a strange experiment and got a strange result:

Boiled S + live R injected into mice → pneumonia → death

This was not expected because boiled S and live R were harmless by themselves

Took blood samples and found live S in the dead mice

Concluded that some factor, a "transforming principle", from the dead S had converted some R bacteria into S bacteria (a genetic change)

Summary of Griffith's experiments

Injected	Result	Live S in Blood?
Live R	No disease	No
Live S	Death	Yes
Killed S	No disease	No
Killed S + Live R	Death	Yes

The Transforming Factor was Found to be DNA

Oswald Avery wanted to know the nature of the transforming principle

Spent many years purifying transforming principle from killed S bacteria

Finally determined it was DNA 1944

First clear cut evidence for hereditary role of DNA

Bacteriophage DNA Changes the Hereditary Functions of Bacteria

Bacteriophage are a type of virus that attacks bacteria

Consist of DNA with a protein coat

When virus attacks bacterium its DNA is inserted, but not its protein

Viruses alter genetic function of bacteria so that they make virus proteins and DNA

Again DNA shown to have a genetic function

BY THE EARLY 1950S IT WAS BELIEVED THAT THE SECRET OF HEREDITY WAS IN THE STRUCTURE OF DNA

Many laboratories began studying the structure of DNA, hoping to find the secret of heredity

The basic composition of DNA was known

A sugar-phosphate backbone, with 4 nucleotide bases (A, C, G & T)

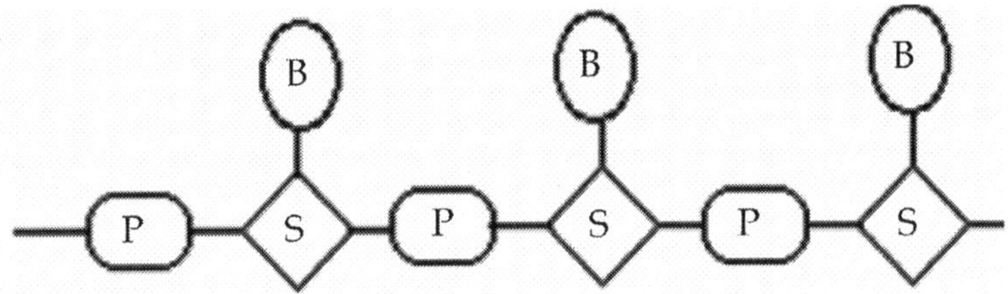

(In the diagram S = sugar; P = phosphate; B = base)

Chargaff had measured the base composition in DNA from many species and found that always A pair with T by two hydrogen bonds and G pair with C by two hydrogen bonds.

X-ray diffraction pictures made by Rosalind Franklin showed that the DNA structure was a helix (2 or more molecules spiraling around each other); the structure appeared to have a uniform thickness

It was found that good H bonds can be formed between bases C & G and also between A & T

Proposed that structure was a double helix held together by H bonds between the bases

The Franklin-Watson-Crick Model of DNA is a Double Helix Held Together by Hydrogen Bonds

The H bonding between C & G and between A & T explains Chargaff's results

They must be equal because every C must be hydrogen-bonded to a G and every A must be hydrogen bonded to a T

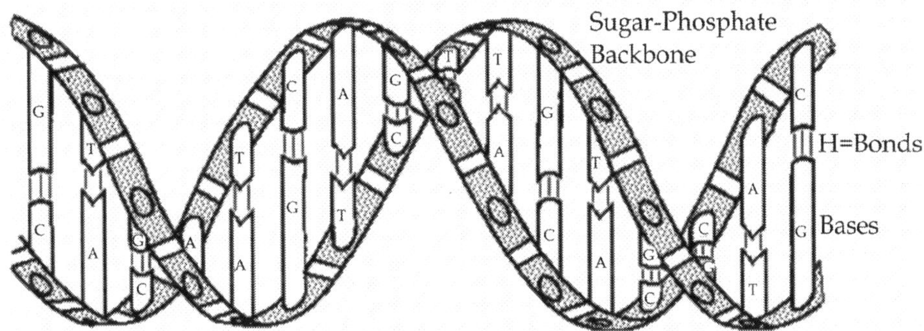

The H bonding of A with T and C with G combines a pyrimidine with a purine (a big one with a little one) and gives the double helix a constant diameter

The 2 DNA chains spiral around each other, taking 10 base pairs to make one rotation

The sugar-phosphate backbone is on the outside of the helix, as Franklin had said it would be

Each DNA Strand is a Template for the Duplication of the Other Strand

The original report of the DNA structure stated: "It has not escaped our notice that the specific pairing we have postulated immediately suggests a possible copying mechanism for the genetic material."

Because of the H bonding requirements each strand of DNA is complementary to the other (they fit together like lock & key)

The copying mechanism: each strand acts as a template to duplicate the other strand

This is called the semiconservative mechaniosm

The copying mechanism was shown with an isotope experiment

Bacteria were grown with N-15 (heavy nitrogen) so that their DNA was made entirely with N-15

The bacteria were then given N-14 (light nitrogen) and the new DNA strands were found to contain half N-14 DNA and half N-15 DNA

After a second duplication of DNA in N-14 half the strands were completely N-14 and the other half were mixed; this is exactly what you would expect if each DNA strand acted as a template

Make a diagram of this experiment and then check your answer

Hydrogen Bonds act as a Zipper Holding the DNA Strands Together

Hydrogen bonds are what make DNA work

Weaker than covalent or ionic bonds

Covalent and ionic bonds ~100 kcal/mole

H bonds ~3 kcal/mole

DNA helix must come apart ("unzip") to duplicate DNA or make RNA (transcription)

Because H bonds are relatively weak, they can be pulled apart without destroying the DNA backbone

Unzipping is local and requires the aid of enzymes

Because they are numerous the H bonds give the DNA helix great stability

DNA can be heated to high temperatures without being destroyed

Helix may come apart, but is easily reformed because of the complementary structure

Samples of DNA in fossils have survived for millions of years with little destruction

DNA is Duplicated by Polymerase Enzymes

Helicase enzymes unwind the helix, starting in an Origin of Replicaqtion site

A primer (RNA) is added to the unwound strand

New DNA is synthesized by DNA polymerases

In the helix the 2 DNA strands run in opposite directions (they are antiparallel)

DNA is duplicated in only one direction: 5' to 3' (the numbers refer to the numbers of carbon atoms on the deoxyribose sugar)

Therefore, the 2 strands are replicated in opposite directions

One strand is duplicated continuously from 5' -> 3'

The other strand is duplicated in fragments (Okazaki fragments), which are later attached (ligated)

CHAPTER - 18

DNA Replication, Transcription and Translation

DNA replication. The double helix is unwound and each strand acts as a template. Bases are matched to synthesize the new partner strands.

DNA replication, the basis for biological inheritance, is a fundamental process occurring in all living organisms to copy their DNA. This process is "replication" in that each strand of the original double-stranded DNA molecule serves as template for the reproduction of the complementary strand. Therefore, following DNA replication, two identical DNA molecules have been produced from a single double-stranded DNA molecule. Cellular proofreading and error toe-checking mechanisms ensure near perfect fidelity for DNA replication.

In a cell, DNA replication begins at specific locations in the genome, called "origins".

Unwinding of DNA at the origin, and synthesis of new strands, forms a replication fork. In addition to DNA polymerase, the enzyme that synthesizes the new DNA by adding nucleotides matched to the template strand, a number of other proteins are associated with the fork and assist in the initiation and continuation of DNA synthesis.

THE CHEMICAL STRUCTURE OF DNA.

DNA usually exists as a double-stranded structure, with both strands coiled together to form the characteristic double-helix. Each single strand of DNA is a chain of four types of nucleotides: adenine, cytosine, guanine, and thymine. A nucleotide is a mono-, di- or triphosphate deoxyribonucleoside; that is, a deoxyribose sugar is attached to one, two or three phosphates. Chemical interaction of these nucleotides forms phosphodiester linkages, creating the phosphate-deoxribose backbone of the DNA double helix with the bases pointing inward. Nucleotides (bases) are matched between strands through hydrogen bonds to form base pairs. Adenine pairs with thymine and cytosine pairs with guanine.

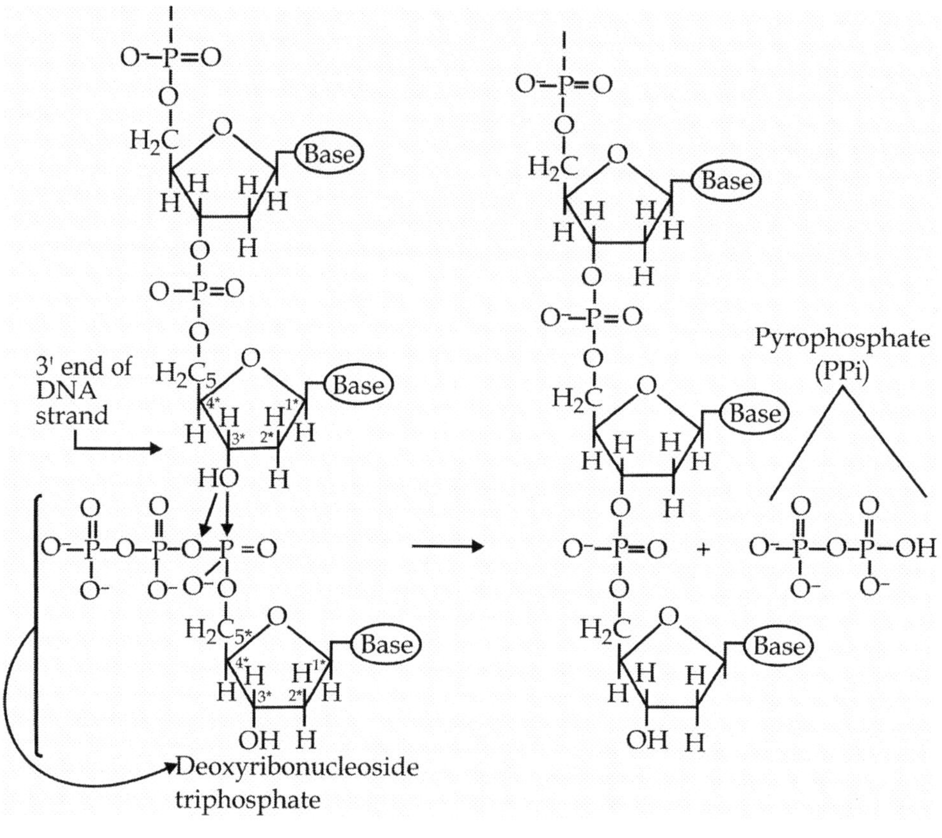

3' end of DNA strand

Pyrophosphate (PPi)

Deoxyribonucleoside triphosphate

DNA strands have a directionality, and the different ends of a single strand are called the "3' (three-prime) end" and the "5' (five-prime) end." These terms refer to the carbon atom in deoxyribose to which the next phosphate in the chain attaches. In addition to being complementary, the two strands of DNA are antiparallel: they are orientated in opposite directions. This directionality has consequences in DNA synthesis, because DNA polymerase can only synthesize DNA in one direction by adding nucleotides to the 3' end of a DNA strand.

The pairing of bases in DNA through hydrogen bonding means that the information contained within each strand is redundant. The nucleotides on a single strand can be used to reconstruct nucleotides on a newly synthesized partner strand.

DNA Replication

Before a cell can divide, it must duplicate all its DNA. In eukaryotes, this occurs during S phase of the cell cycle.

The Biochemical Reactions

- DNA replication begins with the "unzipping" of the parent molecule as the hydrogen bonds between the base pairs are broken.

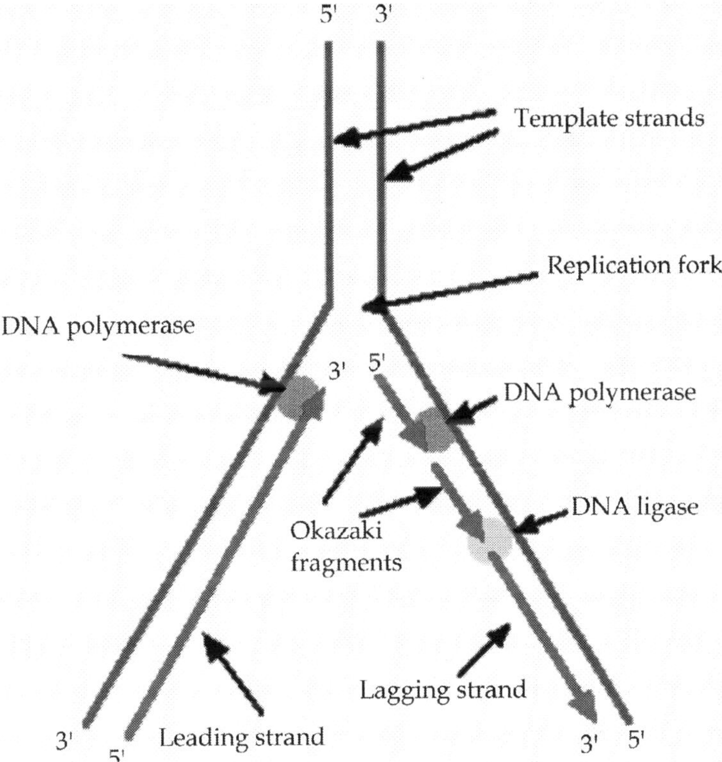

- Once exposed, the sequence of bases on each of the separated strands serves as a template to guide the insertion of a complementary set of bases on the strand being synthesized.
- The new strands are assembled from deoxynucleoside triphosphates.
- Each incoming nucleotide is covalently linked to the "free" 3' carbon atom on the pentose (figure).

- The second and third phosphates are removed together as a molecule of pyrophosphate (**PPi**).
- The nucleotides are assembled in the order that complements the order of bases on the strand serving as the template.
- Thus each C on the template guides the insertion of a G on the new strand, each G a C, and so on.
- When the process is complete, two DNA molecules have been formed identical to each other and to the parent molecule.

The Enzymes

- A portion of the double helix is unwound by a **helicase**.
- A molecule of a **DNA polymerase** binds to one strand of the DNA and begins moving along it in the 3' to 5' direction, using it as a template for assembling a **leading strand** of nucleotides and reforming a double helix. In eukaryotes, this molecule is called DNA polymerase delta (δ).
- Because DNA synthesis can only occur 5' to 3', a molecule of a second type of DNA polymerase (epsilon, å, in eukaryotes) binds to the other template strand as the double helix opens. This molecule must synthesize discontinuous segments of polynucleotides (called Okazaki fragments). Another enzyme, **DNA ligase I** then stitches these together into the **lagging strand**.

DNA Replication is Semiconservative

When the replication process is complete, two DNA molecules — identical to each other and identical to the original — have been produced. Each strand of the original molecule has

- remained intact as it served as the template for the synthesis of
- a complementary strand.

This mode of replication is described as semi-conservative: one-half of each new molecule of DNA is old; one-half new.

Watson and Crick had suggested that this was the way the DNA would turn out to be replicated. Proof of the model came from the experiments of Meselson and Stahl.

Speed of Replication

Bacteria

The single molecule of DNA that is the E. coli genome contains 4.7×10^6 nucleotide pairs. DNA replication begins at a single, fixed location in this molecule, the **replication origin**, proceeds at about 1000 nucleotides per second, and thus is done in no more than 40 minutes. And thanks to the precision of the process (which includes a "proof-reading" function), the job is done with only about one incorrect nucleotide for every 10^9 nucleotides inserted. In other words, more often than not, the E. coli genome (4.7×10^6) is copied without error!

Eukaryotes

The average human chromosome contains 150×10^6 nucleotide pairs which are copied at about 50 base pairs per second. The process would take a month (rather than the hour it actually does) but for the fact that there are many places on the eukaryotic chromosome where replication can begin. Replication begins at some replication origins earlier in S phase than at others, but the process is completed for all by the end of S phase. As replication nears completion, "bubbles" of newly replicated DNA meet and fuse, finally forming two new molecules.

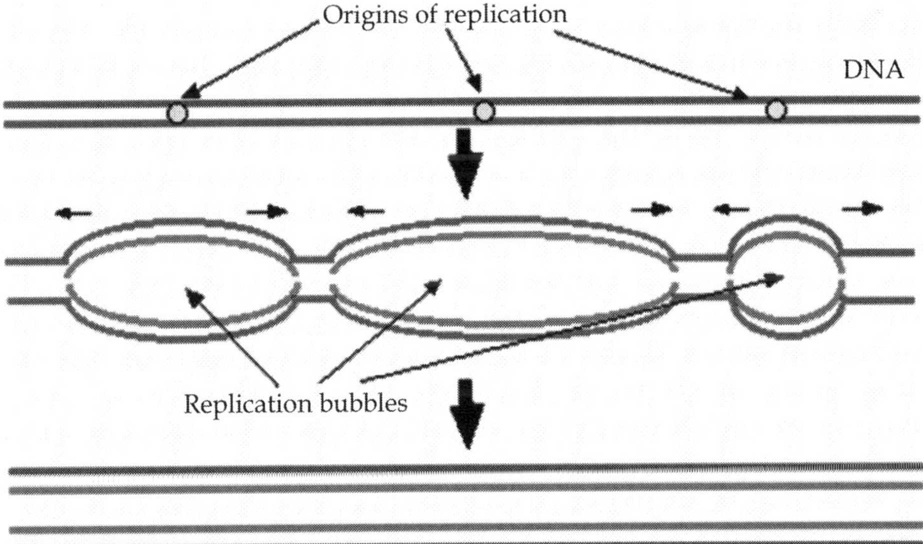

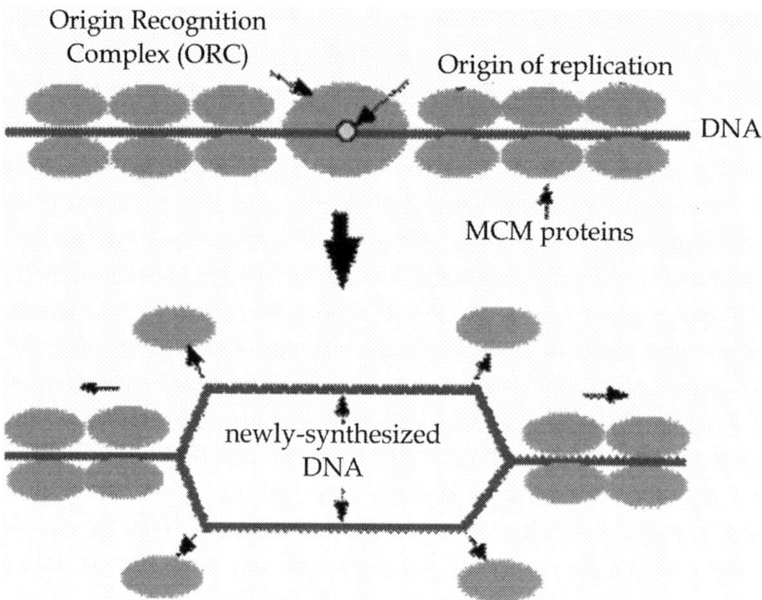

Control of Replication

With their multiple origins, how does the **eukaryotic cell** know which origins have been already replicated and which still await replication?

An observation: When a cell in G_2 of the cell cycle is fused with a cell in S phase, the DNA of the G_2 nucleus does not begin replicating again even though replication is proceeding normally in the S-phase nucleus. Not until mitosis is completed, can freshly-synthesized DNA be replicated again.

Two control mechanisms have been identified – one **positive** and one **negative**. This redundancy probably reflects the crucial importance of precise replication to the integrity of the genome.

Licensing: positive control of replication

In order to be replicated, each origin of replication must be bound by:

- an **O**rigin **R**ecognition **C**omplex of proteins (**ORC**). These remain on the DNA throughout the process.
- Accessory proteins called **licensing factors**. These accumulate in the nucleus during G_1 of the cell cycle. They include:
    - Cdc-6 and Cdt-1, which bind to the **ORC** and are essential for coating the DNA with
    - **MCM proteins**. Only DNA coated with MCM proteins (there are 6 of them) can be replicated.

Once replication begins in S phase,

- Cdc-6 and Cdt-1 leave the ORCs (the latter by ubiquination and destruction in proteasomes).
- The MCM proteins leave in front of the advancing replication fork.

Geminin: negative control of replication

G_2 nuclei also contain at least one protein — called **geminin** — that prevents assembly of MCM proteins on freshly-synthesized DNA (probably by blocking the actions of Cdt1).

As the cell **completes mitosis**, geminin is degraded so the DNA of the two daughter cells will be able to respond to licensing factors and be able to replicate their DNA at the next S phase.

Some cells deliberately cut the cell cycle short allowing repeated S phases without completing mitosis and/or cytokinesis. This is called **endoreplication**. How these cells regulate the factors that normally prevent DNA replication if mitosis has not occurred is still being studied.

Protein Synthesis: Transcription

The genes tell the cell how to make proteins via transcription

- They coined the phrase "One gene, one enzyme."
- By controlling enzymes genes control what kind of chemistry goes on in cells
- Not all proteins are enzymes- some are structural

The genes have a triplet code

- Genes must have a code specifying how proteins should be made
- Nature uses 3 base words or "codons" for the genetic code
    - Some amino acids have more than 1 codon
    - A gene must have a starting point and an ending point
        - 3 codons are stop codons (ATT, ATC and ACT)
        - There is 1 start codon, which also codes for the amino acid methionine (TAC)
- The genetic code was determined by using artificial RNA molecules (see below) to see what proteins they coded for

The genetic code is universal

- Apparently, all species on earth use the same genetic code
- A pea plant can translate human genes and vice versa
- This points to a common origin of life on earth
- The universal code is very useful in the biotech industry- we can make human proteins in other animals and plants
- Some minor exceptions to the universal code are found in mitochondrial & chloroplast DNA and in a few protozoa

The genetic code

First base	Second base				Third base
	U	C	A	G	
U	Phenylalanine	Serine	Tyrosine	Cysteine	U
	Phenylalanine	Serine	Tyrosine	Cysteine	C
	Leucine	Serine	Stop (Ochre)	Stop (Opal)	A
	Leucine	Serine	Stop (Amber)	Tryptophan	G
C	Leucine	Proline	Histidine	Arginine	U
	Leucine	Proline	Histidine	Arginine	C
	Leucine	Proline	Glutamine	Arginine	A
	Leucine	Proline	Glutamine	Arginine	G
A	Isoleucine	Threonine	Asparagine	Serine	U
	Isoleucine	Threonine	Asparagine	Serine	C
	Isoleucine	Threonine	Lysine	Arginine	A
	Methionine (start)	Threonine	Lysine	Arginine	G
G	Valine	Alanine	Aspartic acid	Glycine	U
	Valine	Alanine	Aspartic acid	Glycine	C
	Valine	Alanine	Glutamic acid	Glycine	A

When proteins are made a "working copy" of the gene is made from RNA

- DNA does not leave the nucleus, but proteins are made on ribosomes in the cytosol
- The code message must be sent from the nucleus to the cytosol
- When a protein is to be made a working copy of the gene is made out of RNA; this is called transcription

- The DNA strands act as templates for making RNA copies of genes
- The RNA copy is complementary to the DNA copy and the base T is replaced by U

  -T-C-A-T-T-G-T-G-C-A-A-C- DNA gene

  -A-G-U-A-A-C-A-C-G-U-U-G- m-RNA copy
- The RNA copy of the code is called messenger RNA or m-RNA
- This arrangement keeps the DNA master copy of the gene in the nucleus where it is protected
- The working copy in RNA is made only when it is needed and is usually rapidly broken down

Messenger RNA is made by RNA polymerase (Transcription)

- The RNA copy of DNA is made by an enzyme, RNA polymerase
- Only the coding strand of DNA is transcribed
- RNA polymerase binds to promoter region, just upstream from the coding section of the gene (TATA box)
- Protein transcription factors aid the binding of RNA polymerase, control RNA synthesis
- Gene is transcribed:
  - From an initiation site (about 25 bases downstream from the polymerase binding site)
  - To a termination site (commonly AATAAA)
  - **Transfer of genetic information is (**Transcription is from the 5' to the 3' end almost)

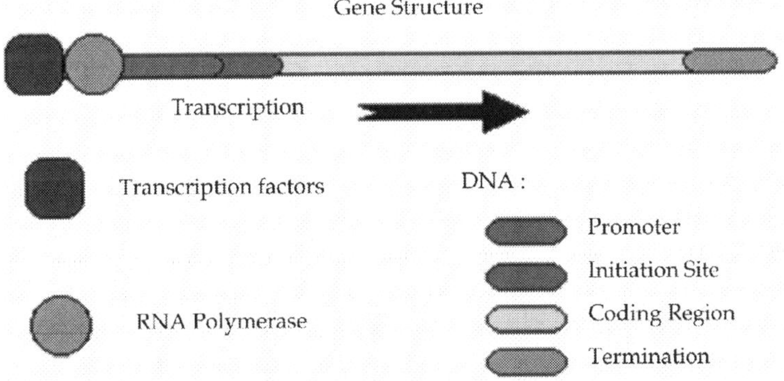

Always: → DNA → RNA → Protein

- Genetic information is stored as DNA, transcribed to RNA and translated to protein:
  - DNA → RNA → protein
- Information does not flow in the reverse direction
  - One exception: retroviruses store genetic information as RNA
  - When they infect a cell they make a DNA copy of their RNA, using an enzyme, reverse transcriptase
  - One exception: retroviruses store genetic information as RNA
  - When they infect a cell they make a DNA copy of their RNA, using an enzyme, reverse transcriptase

Genes Have 2 Types of DNA: Exons and Introns

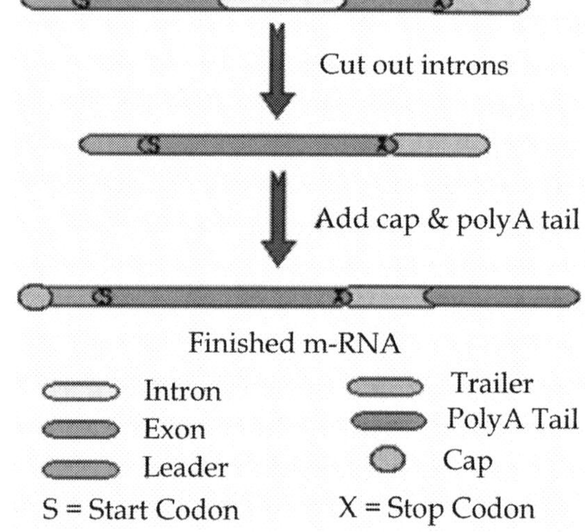

- Eukaryotic cells have split genes
- Non-coding sequences called introns are inserted in the middle of genes
- Prokaryotic genes seldom have introns: they are not split
- Prokaryotes have less room to store their genetic information
- Coding sections are called exons
- Both types of DNA are transcribed into m-RNA
- Later the introns are cut out by complexes called spliceosomes
- Exons are ligated together
- The messenger RNA is also given a cap and a polyadenine tail before being sent to the cytosol
- GTP cap may help attach m-RNA to ribosome
- A long polyA tail increases the half life of the m-RNA

Synthesis and Destruction of RNA is Used to Control Metabolism in Cells

- Messenger RNA controls the rate at which proteins are synthesized in cells
- To reduce the amount of an enzyme a cell can decrease the rate of m-RNA transcription for its gene
- Alternatively the cell can increase the rate at which the m-RNA is destroyed

Protein Synthesis: Translation

The genetic code is translated on the ribosomes

Translation consists of 2 simultaneous actions:

Deciphering the genetic code

Formation of a protein according to the directions given by the code

Translation takes place outside the nucleus, on the ribosome

Two types of ribosomes:

Free ribosomes

Ribosomes attached to the endoplasmic reticulum (ER)- forms the rough ER

All protein synthesis starts on free ribosomes

If the growing protein has a certain "signal sequence" the ribosome attaches to the ER

Proteins made on the ER are destined to be secreted or routed to certain organelles such as the cell membrane or lysosomes

Proteins made on free ribosomes go to the cytosol, mitochondria and chloroplasts

Three Types of RNA are Involved in Synthesizing Proteins

Three types of RNA (all formed in nucleus) are used in translation:

Messenger RNA (m-RNA): carries working copy of the code for a protein

Ribosomal RNA (r-RNA): furnished structure and function for ribosomes (60% RNA): r-RNA includes the enzyme, peptidyl transferase

Transfer RNA (t-RNA): carries individual amino acids from the cytosol to the ribosome

At least 20 types needed because there are 20 different types of amino acids

The Ribosomes Have a Large and a Small Subunit

Ribosomes are made in the nucleolus

Each ribosome has a large unit and a small unit

Separate unless making protein

In eukaryotic cells subunits are designated 60S and 40S (S = sedimentation coefficient; large subunit has large number, sediments faster in ultracentrifuge)

Prokaryotic cells have smaller subunits: 50S and 30S

In protein synthesis the small subunit binds to the m-RNA and then to the large subunit to make the complete ribosome

Details of the ribosome assembly:

When the small subunit binds the m-RNA it moves along it to find the start codon (AUG)

The start codon is also used to code methionine; every eukaryotic protein starts with methionine (in some cases the amino acid is later removed)

Then a t-RNA molecule carrying methionine binds to the AUG codon

This causes the large subunit to bind to the complex

Many ribosomes can attach to and read the same m-RNA

Produces polyribosomes

A Gene is Located in an Open Reading Frame on the m-RNA

A gene is a sequence of codons between a start codon and a stop codon

This structure is called an open reading frame, which means that a ribosome can read it without being stopped

There can be no stop codons in the middle of a gene

A long open reading frame will not be formed randomly very often because 3 of the 64 codons are stop codons (about 5%)

The Ribosome Reads the Message, One Codon at a Time

The ribosome positions itself over the start codon and then moves along the messenger RNA 3 bases at a time, from codon to codon

Movement requires energy from guanosine triphosphate (GTP)- similar to ATP

Movement is always from the 5' direction to the 3' direction

Transfer RNA Matches Amino Acids to Codons

Amino acids are shuttled to the ribosome by 20 types of transfer RNA (t-RNA)

Small RNA molecules, cloverleaf structure

Each t-RNA has a specific amino acid binding site and an anticodon site

Binding of amino acids requires ATP and a special enzyme (20 types of enzyme- one for each amino acid)

Anticodon site is complementary to the codon on m-RNA

Example: if the codon on m-RNA is CUA the t-RNA anticodon will be GAU

Assures that correct amino acid is attached for each codon: the CUA/ GAU codon/anticodon pair will cause the amino acid leucine to be added to the protein

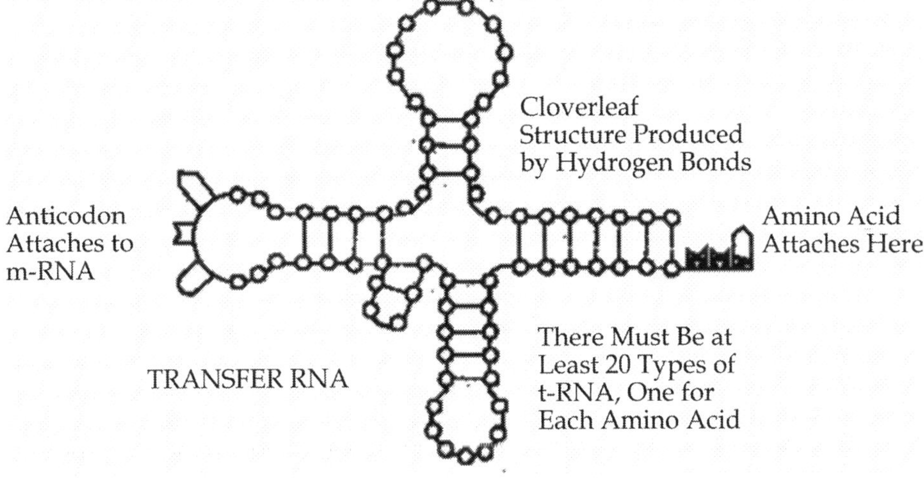

Amino Acids are Added to the Protein One at a Time

Ribosome (small subunit) has 2 binding sites for t-RNAs to bring in amino acids

P site has the peptide, attached to the last t-RNA

A site has the next amino acid attached to its t-RNA molecule

Binding site is where t-RNA anticodon binds to codon on the m-RNA

Steps in protein synthesis:

Start codon (AUG) is exposed at the P binding site and binds to the t-RNA for methionine

The next codon is exposed at the A site

A t-RNA with the correct anticodon binds to the A site; there are now 2 amino acids attached to the ribosome- one at the P site and one at the A site

The 2 amino acids are attached together, forming a peptide bond (this reaction is catalyzed by peptidyl transferase, an RNA enzyme); the peptide is attached to the A site t-RNA

The P site t-RNA leaves the ribosome

The A site t-RNA, with its peptide is transferred to the P site and the ribosome shifts 1 codon along the m-RNA

This exposes the next codon at the A site

The process of adding amino acids is repeated over and over until the stop codon is reached

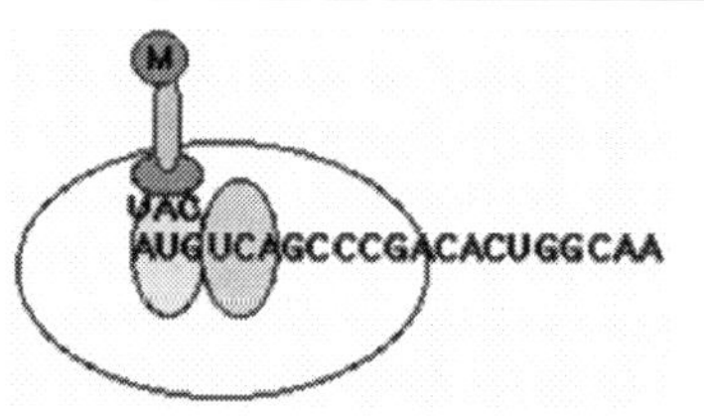

Initial steps: Messenger RNA is bound to ribosome with the start codon (AUG) at the P site. transfer RNA molecule with the amino acid methionine (M) and the anticodon UAC has bound to the exposed start codon. The codon UCA is exposed at the A site.

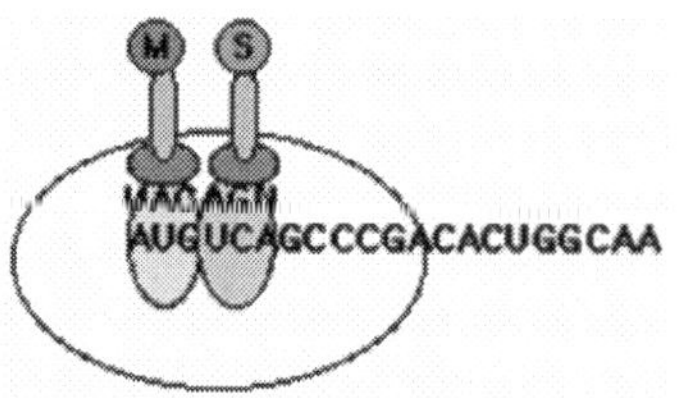

A second transfer RNA molecule, with the anticodon AGU and the amino acid serine (S) has bound to the A site.

Contd...

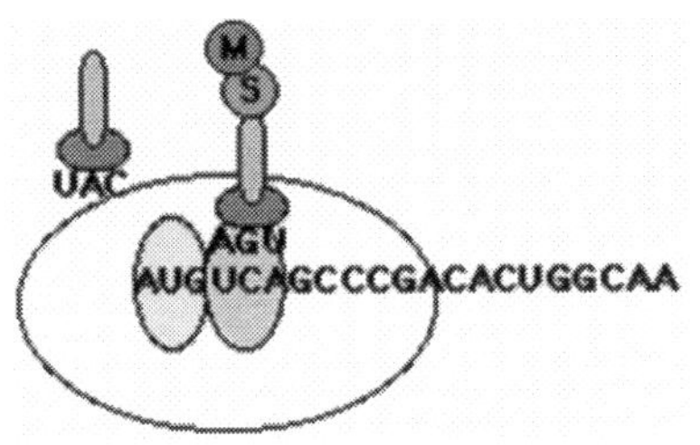

A peptide bond has formed between M and S and the peptide is bound to the A site. The methionine transfer RNA leaves, and the P site is exposed.

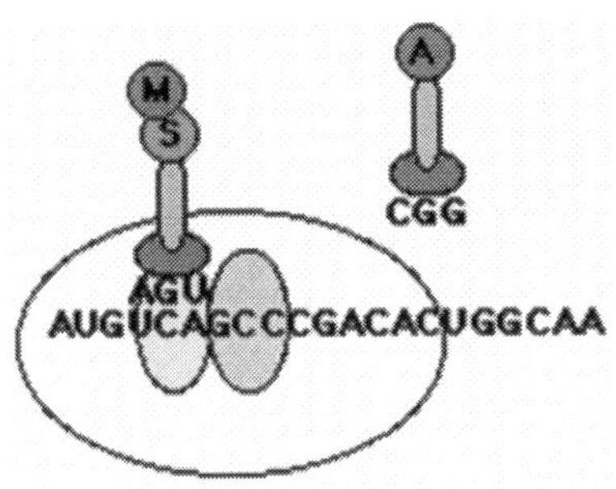

The ribosome has moved along the messenger RNA one codon, bringing the peptide to the P site. This exposes the A site and the next transfer RNA, carrying alanine (A) is about to bind.

Review of the Coding: DNA → RNA → Protein

	Start	1st Codon	2nd Codon	3rd Codon
DNA code	TAC	AGT	CGG	GCT
m-RNA code	AUG	UCA	GCC	CGA
t-RNA anticodon	UAC	AGU	CGG	GCU
Amino Acid	Methionine	Serine	Alanine	Arginine

Energy cost for protein synthesis

- Charging of tRNA : 2 ATP's
- Binding of tRNA to Ribosome : 1 GTP
- Translocation : 1 GTP
- Total Cost : 4 High Energy Phosphate Bonds for each Peptide Bond Formation

Protein Maturation

A mutation is an inherited change in the genetic code of a gene

Changing the DNA bases will change the protein code

In many cases this will be lethal and the cell will die

The changes may alter the active site of the protein

Stop codons may be inserted so that most of the gene cannot be translated

The protein may still work, but may have harmful side effects

If DNA changes are passed on to the next generation the changes are called mutations

Mutations occur naturally due to attack of the DNA by chemicals, radiation, etc.

Most potential mutations are repaired, but some are not

Natural selection eliminates harmful mutations

In the absence of selection mutations will build up in genes; if gene is not needed it will be wrecked

Stop codon mutations will tend to destroy genes not subject to selection pressure

Example: gene necessary for the production of vitamin C has been wrecked in many vertebrates including humans; we now must get vitamin C from the diet

The damaged enzyme is called gulonolactone oxidase; the gene is located on human chromosome 8

Gene is damaged in primates, guinea pig, Indian fruit bat and some birds

The simplest type of mutation is a point mutation; a single DNA base is changed to another base

Example of point mutation: sickle cell anemia

A defect in hemoglobin, the protein that carries oxygen in the blood

A point mutation in which a single amino acid in one of the hemoglobin chains has been altered

A glutamic acid is changed to a valine in the number 6 position of the beta chain

This causes the hemoglobin to crystallize and damage the red blood cells

Red cells destroyed → anemia

Red cells plug capillaries → tissue damage

The codons for glutamic acid are: GAA and GAG

Codons for valine are: GUA, GUC, GUG and GUU

What base changes are the most likely cause of the sickle mutation?

The sickle gene is high in malaria areas of the world (Africa, Southeast Asia) because those who are heterozygous (one A gene and one S gene) are partially protected against malaria

A = normal hemoglobin gene; S = sickle hemoglobin gene

Those with 2 S genes have severe anemia

Current treatment: hydroxyurea, turns on production of fetal hemoglobin which relieves symptoms

Possible gene therapy

CHAPTER - 19

Genetics of Viruses

Viruses are much simpler than cells

Protein capsid

Genetic information stored on DNA or RNA:

- Types of genome:
  - Double strand DNA: sometimes circular
  - Single strand DNA: sometimes circular
  - Double strand RNA
  - Single strand RNA

Some can be crystallized

Viruses are Intracellular Parasites

Should viruses be considered living organisms?

Viruses have no metabolic enzymes or organelles

Cannot live independently; can reproduce only within a host cell

- In reproduction virus supplies a few genes
- Host supplies ATP, amino acids, nucleotides, enzymes for duplicating genes, enzymes for making proteins, ribosomes,t-RNA, etc

Viruses Contain only a Few Genes

Viruses have few genes

- Some have only 4 genes

Bacteriophages have as many as 300 genes

Carry genes for capsid proteins

Also have some genes required for replication (such as reverse transcriptase in retroviruses)

Bacteriophages Attack Bacteria

Bacteriophages are viruses that attack bacteria

They are most complex viruses

DNA viruses, linear, not circular

Have polyhedral head + cylindrical tail

Attach to bacterial surface and inject their DNA into the bacterium; protein does not enter

Phage Have Lytic and Lysogenic Lifestyles

After the DNA is injected into the bacterium a circular DNA molecule is produced

At this point the virus can move into either the lytic or the lysogenic pathway

The lytic pathway:

The phage DNA is transcribed and translated to make phage proteins

The phage DNA is duplicated; sometimes the host DNA is broken down by virus nucleases to get the building blocks needed to make virus DNA

The new virus proteins and DNA assemble themselves into new viruses

The new viruses cause the host bacterium to lyse, releasing the viruses to infect other cells

Note: In animal viruses the lytic cycle does not always kill the host cell

The lysogenic pathway:

The virus DNA becomes intergrated into the bacterial chromosome; in this state the virus is called a provirus

Each time the bacterium divides the virus DNA is duplicated along with the bacterial chromosome

Environmental stimuli (radiation, UV) cause the virus DNA to be released from the bacterial chromosome

The virus DNA then enters the lytic cycle, and is eventually released when the cell lyses

Retroviruses Can Make DNA from RNA

Retroviruses have single strand RNA (ssRNA) as their genetic material

Infect animal cells

These viruses inject the enzyme reverse transcriptase along with the RNA

Reverse transcriptase makes double stranded DNA from the single strand RNA

First makes DNA complementary to the RNA → DNA/RNA hybrid, hydrogen bonded together

Then hydrolyzes the RNA → single strand DNA (ssDNA)

Makes a complementary copy of the DNA → double strand DNA (dsDNA)

The virus DNA inserts into the animal cell DNA

Virus DNA is transcribed & translated to make capsid proteins & reverse transcriptase

Some of the virus RNA is put into the new capsids to form new virus

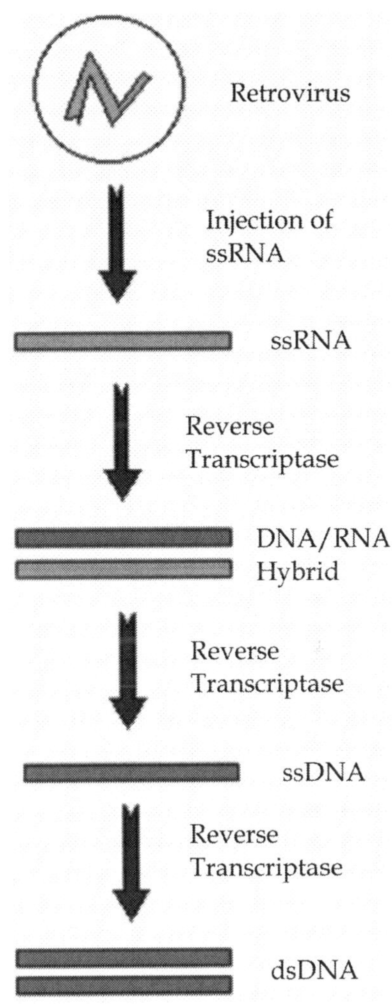

Viruses Cause Many Human Diseases

Viruses cause many diseases in animals and plants

Some human virus diseases are Cancers (Cervical, Leukemia, Burkitt's lymphoma), Common cold, Smallpox, Polio, Measles, Rabies, Influenza, AIDS,

Types of virus damage are Lysis of cells, Production of virus toxins, Stimulation of production of toxins by bacteria, Transformation into tumor cell, Overstimulation of body defense systems

AIDS is Caused by a Retrovirus

AIDS is a human disease in which the immune system is badly damage

Transmitted by body fluids

Semen: unprotected sex

US & Europe: primarily homosexual

Africa & Asia: primarily heterosexual

Blood: drug users, transfusions

Breast milk

Virus invades and destroys helper T-cells

Viral surface glycoprotein (protein with sugar groups) binds to CD4 protein of helper T cell:

Virus enters T cell

Virus coat is removed inside the cell

Virus core enters the host cell nucleus

Reverse transcriptase converts RNA genes to dsDNA

Virus DNA is inserted into a host chromosome by integrase enzyme

Provirus stage

Virus has long lysogenic phase; ~10 year average

Produces virus RNA while in the provirus state

Virus RNA used to make new virus which bud from the host cell membrane

Gradually destroys helper T cells

Person with reduced helper T cells becomes susceptible to opportunistic infections

Kaposi's sarcoma, Pneumocystis pneumonia

CHAPTER - 20

Genetics of Bacteria

BACTERIA HAVE SMALL, COMPACT GENOMES

We now know the complete DNA sequences for about a dozen bacteria

Bacteria seem to need about 500 to 5000 genes to conduct their lives

Because they are small, bacterial genomes are highly organized

Most of the time there are only a few bases between the end of one gene and the beginning of the next

Few introns are found in bacteria

Bacteria contain accessory "chromosomes" called plasmids; each has only a few genes

Feedback Mechanisms Control Transcription of Bacterial Genes

Genes have regulatory sections of DNA

Many bacterial genes are organized into operon units

Promoter region which binds RNA polymerase

Regulatory gene which makes a repressor molecule

Operator unit which switches the gene off when it binds the repressor (blocks binding of RNA polymerase)

Several related genes may be controlled by the same promoter (polycistronic)

They are transcribed together

Example: Lac operon of *E. coli*

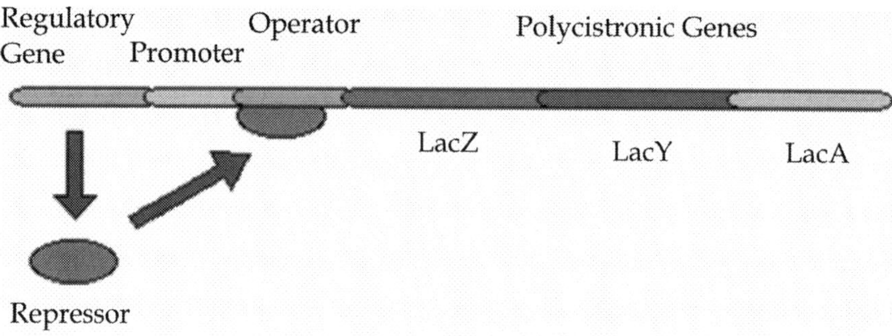

This operon is switched off because the repressor molecule is bound to the operator

This is the usual state- the bacterium does not need these enzymes unless there is lactose in the medium

If the bacterium detects lactose small amounts leak into the cell, setting in motion the events that will turn on the lac operon

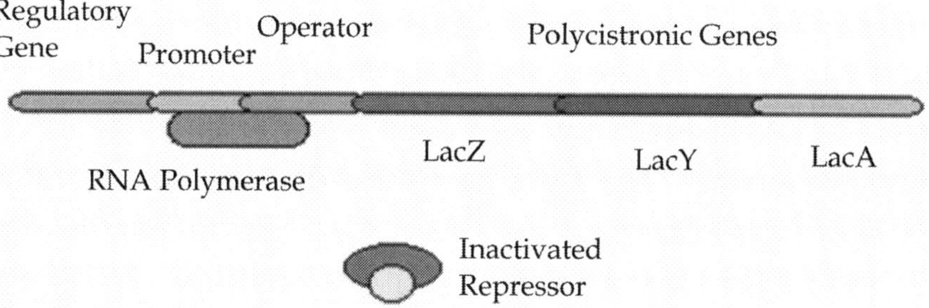

Some of the lactose is converted to a molecule which binds the repressor, inactivating it and pulling it off of the operator site

This switches the operon on, allowing it to bind RNA polymerase and transcribe

The cell then makes the enzymes needed to metabolize lactose (a sugar)

This process is called induction

Bacteria Have 3 Ways of Making Recombinant DNA

Recombinant DNA is the fusion of DNA from 2 separate individuals

There are 3 ways of transferring DNA into bacterial cells

Transformation

Transduction

Conjugation

Detection of recombination:

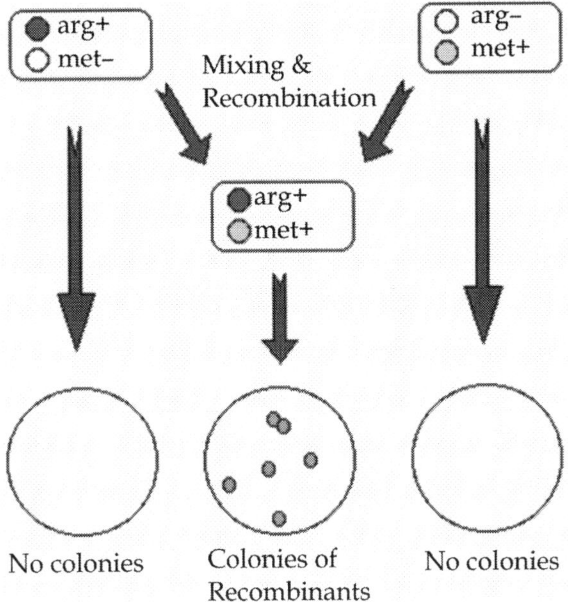

Two strains of mutant bacteria cannot make essential amino acids (arg- & met-) and therefore cannot grow

When mixed a small fraction recombines to form bacteria that can make all essential amino acids

Bacteria Can Get Genes from Naked DNA: Transformation

Sometimes naked DNA is taken up by bacterial cells

This is how the transforming principle DNA got into the pneumonia bacteria in Griffith's experiments

Some bacteria have specialized membrane proteins to bring DNA into their cells

Ca stimulates uptake of DNA into bacteria

Phage Can Carry Genes from One Bacterium to Another: Transduction

Bacteriophage infecting a bacterium can package some bacterial DNA into capsids

When capsid infects a new bacterium it can transfer the bacterial genes to the new individual

Two ways of picking up bacterial DNA:

Generalized

Phage hydrolyzes host DNA

Chunks of host DNA are accidentally packaged into capsids

Specialized:

Bacteriophage insert into bacterial chromosome (provirus state)

When virus leaves the provirus state it may take some bacterial genes close to the insertion sitewith it

Bacteria Can Swap Genes through Sex Pili: Conjugation

Bacteria have a form of "sex" called conjugation

Not related to reproduction

Usually only a few genes transferred

"Male" donor bacterium projects a hollow tubules called pili

Male bacterium has an F plasmid

F plasmid has genes needed to make the sex pili

F plasmid DNA inserts into bacterial chromosome; converts bacterium into Hfr (high frequency of recombination) cell

DNA travels in only one direction

Pili connect to a "female" receptor cell

"Female" bacterium has no F plasmid

"Female" can take up an F plasmid and become "male"

Donor does not lose DNA; his chromosome is duplicating while he transfers genes

Usually only part of the chromosome is transferred before the bacteria break apart

Transfer of genes is slow; can be used to map bacterial genes

Receptor female becomes partially diploid

Plasmids Allow Bacteria to Share Genes Rapidly

Plasmids are small circular accessory "chromosomes"

Not necessary for life, but give special properties

Duplicate independently of bacterial chromosome

Some can insert into the bacterial chromosome

Episome can reproduce when inserted or when free

Examples:

F plasmid: makes bacterium "male"

R plasmid: gives antibiotic resistance

Plasmids are like viruses, but have no etracellular phase

Restriction Enzymes Protect Bacteria from Virus Infections

Certain bacteria have enzymes which cut DNA at very specific sites

These enzymes are called restriction enzymes

Used to defend bacteria against virus DNA

Foreign DNA is cut up

Bacterium protects its own DNA by methylating DNA bases

Restriction enzymes cut at specific sequences, 4 to 8 bases long

Restriction sites are palindromes- the sequence on the complementary strand is the same when read backwards:

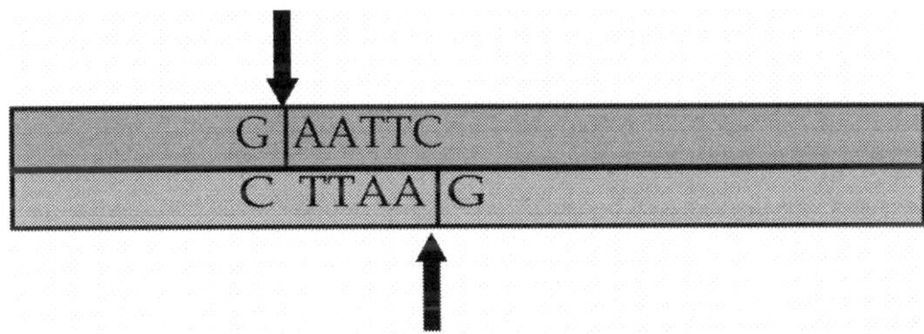

A cut is made in each strand so that the molecule separates; arrows (above) show where molecule is cut

The overlaping ends make the DNA strands "sticky": they can bind to other strands which have complementary sequence

Uses of restriction enzymes:

Making recombinant DNA

Cutting DNA into manageable fragments for sequencing

Mapping of chromosomes

Comparing Bacterial & Virus Genomes with Those of Higher Organisms

Size: bacterial & virus genomes much smaller than those of higher organisms

Organization: bacterial & virus genomes very compact: few introns or "nonsense" DNA

Nematode DNA: averages 5 introns/gene

Only ~ 10% of mammalian DNA codes for proteins

Gene organization: bacterial polycistronic; in higher organisms genes are usually transcribed independently

Bacteria mostly haploid; higher organisms mostly diploid- back up copy of genes

Adaptation: bacteria can make rapid changes in their population to adjust to the environment; in higher organisms individuals adapt by enzyme induction

Table gives data for 6 organisms that have been completely sequenced

Data for humans is given for comparison

Species	Number of Chromosomes	Millions of Base Pairs	Number of genes
AIDS virus	1 H	0.009	15
Mycoplasma genitalium (small bacterium)	1 H	0.6	390
Methanococcus jannaschii (archaeon bacterium)	1 H	1.7	1,738
Escherichia coli(bacterium)	1 H	4.6	4, 289
Saccharomyces cerevisiae (yeast)	16 H	12.1	6, 217
Caenorhabditis elegans (nematode worm)	11 (male) D12 (female) D	97	19, 099
Homo sapiens (humans	46 D	3000	100,000 (estimate)

H = haploid; D = diploid

Organisms:

The AIDS virus is a retrovirus, with its genome encoded in RNA

Mycoplasma genitalium has the smallest genome known for any free-living creature.

The autotroph (makes its own food), Methanococcus jannaschii, has 1738 genes

M. jannaschii lives near hydrothermal vents, 3000 meters beneath the sea; its preferred temperatures are 49 to 94 deg C

Escherichia coli is the common intestinal bacterium widely used in research

Saccharomyces cerevisiae is a yeast, a single eukaryotic cell

Caenorhabditis elegans is a nematode worm with exactly 959 cells which lives in the soil. It is widely used in developmental research. It has the largest genome which has been completely sequenced. C. elegans does not have true females, only males (with 11 chromosomes) and hermaphrodites (with 12)

The gene figure for humans is an estimate

CHAPTER - 21

Recombinant DNA Technology

Restriction enzymes cut only at specific sequences of 4 to 8 bases

A long restriction site is rare. Cutting rare sites will produce a few large pieces of DNA.

Restriction enzymes with long sites → small number of large fragments

Restriction enzymes with short sites → large number of small fragments

Number of bases	Probability	Average Number of Fragments for1 Million Bases	Average Size of Fragment (Base Pairs)
4	1/256	3907	256
5	1/1024	978	1024
6	1/4096	245	4096
7	1/16384	62	16384
8	1/65536	16	65536

There are now about 200 known restriction enzymes

Some uses for restriction enzymes:

Sequencing DNA: restriction enzymes cut different copies of DNA at exactly the same spot

Large amounts of the same fragment can be obtained for analysis

Cloning DNA: restriction enzymes allow formation of recombinant DNA

Fragment Size Can be Analyzed by Gel Electrophoresis

DNA has a negative charge, and if placed in an electric field will migrate toward the positive pole

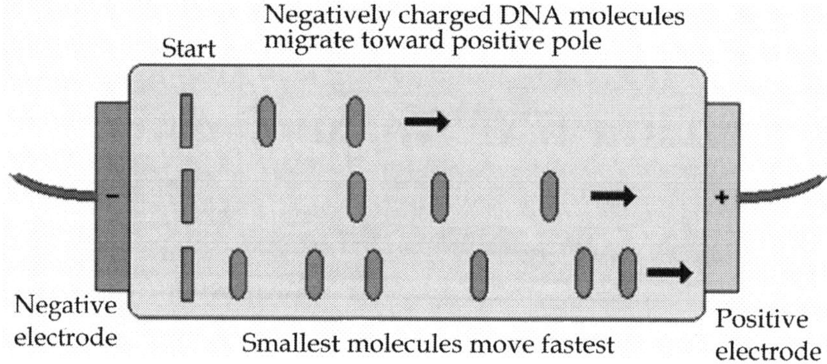

Small fragments will move faster than large ones, according to a logarithmic relationship, in agarose gels

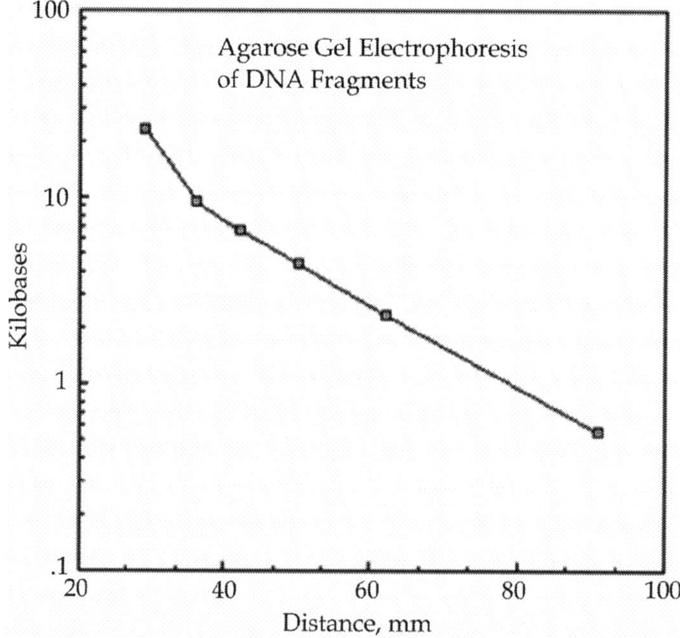

This can be used to determine the sizes of DNA fragments in experiments

Methods to make the DNA visible:

Autoradiography: radioactive DNA; gel exposed to a photographic plate to make bands visible

Stain with ethidium bromide: fluorescent dye- must be handled carefully because it is carcinogenic

Stain with methylene blue- not as sensitive as ethidium, but safer

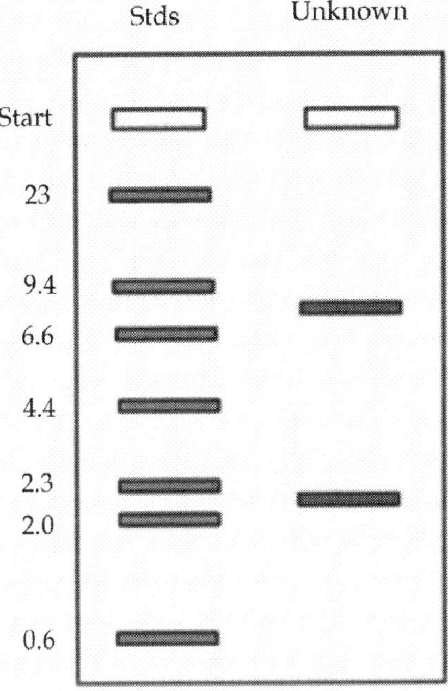

Example:

The example shows calibration standards, running from 0.6 to 23 kilobases, on the left

The unknown has 2 bands, of approximately 7.5 and 2.1 kilobases

Restriction Enzymes Can be Used to Make Recombinant DNA

If a plasmid and the DNA sample of interest have some of the same restriction sites the DNA can be inserted into the plasmid

Steps in making recombinant DNA:

Cut both the plasmid and the DNA sample with the same restriction enzyme

This will produce "sticky ends" on both plasmid and DNA sample

The sticky ends are complementary and will hydrogen bond to each other

Make the attachment permanent with the enzyme DNA ligase (catalyzes formation of phosphodiester bonds)

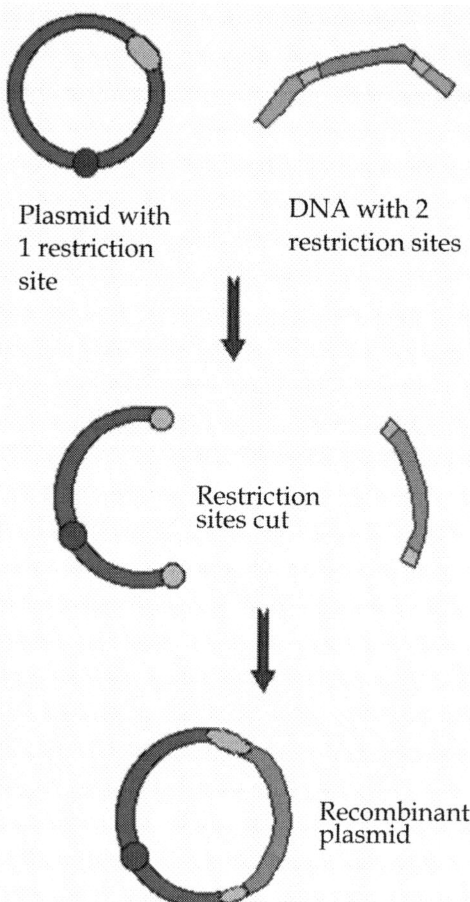

The dot on the plasmids represents the ampicillin-resistance gene

Similar methods can be used to make recombinant DNA with bacteriophage

Recombinant DNA was first made by these methods by Herbert Boyer (Stanford University) and Stanley Cohen (UC, San Francisco) in 1973

Several Tricks are Used to Detect Recombinants

The recombinant plasmids are put into bacteria so that multiple copies can be made

When the bacteria divide the plasmids divide too

If bacteria are diluted and grown on agar plates they will form round colonies

Each colony is a clone derived from a single cell

Tricks are used to find colonies with DNA inserts

Trick 1: plasmids with a gene that gives resistance to the antibiotic ampicillin are used

If the bacteria are grown on agar plates containing ampicillin only bacteria with a plasmid can grow

Trick 2: DNA is inserted at a restriction site in the middle of a beta-galactosidase gene (Lac Z)

If DNA is inserted the gene is destroyed

Bacteria with beta-galactosidase gene can convert an artificial substrate, X-gal, into a blue dye

If the colonies are given X-gal those without recombinant DNA will be blue (their gene works) and those with recombinant DNA will be white (their gene has been destroyed)

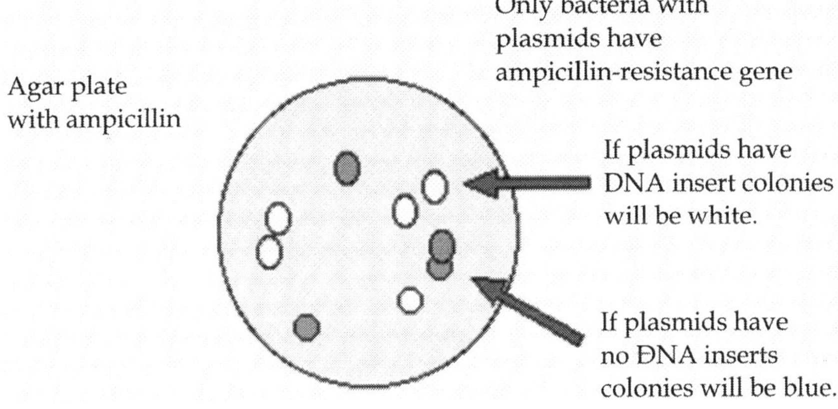

The Nuclear DNA Can be Cut up to Make a Genomic DNA Library

If all of the DNA in the nucleus is cut up with restriction enzymes and made into plasmid recombinant DNA thousands of different fragments will be obtained

This is called a genomic DNA library since it contains all of the DNA from the genome

Bacteria will not be able to make proteins from a genomic library because they do not know how to cut out introns

If proteins are required the DNA can be grown in eukaryotic cells

Messenger RNA Can be Used to Make a Complementary DNA Library

Another way of making a DNA library is to start with messenger RNA

The m-RNA can be converted to double strand DNA using the enzyme, reverse transcriptase

The DNA can then be put into plasmids to make a complementary DNA library

This type of library shows the genes that are actually being transcribed into m-RNA

Bacteria can make proteins from a cDNA library because there are no introns

CHAPTER - 22

Genetic Engineering: PCR, RFLP Analysis and Gene Therapy

The polymerase chain reaction (PCR) is a rapid way of amplifying (duplicating) specific DNA sequences

Method was devised by Kary Mullis of Cetus Corporation, Emeryville

He recieved a $20,000 bonus and later a Nobel Prize

DNA heated to high temperature is not destroyed; separates into single strands, but reforms helix when cooled

PCR Method:

DNA to be amplified is put into solution containing:

Short DNA "primers" which can bind to the 3' ends of the DNA

The 4 nucleotide bases: A, C, G, T

DNA polymerase

Special DNA polymerase (Taq polymerase) working at very high temperature is used; isolated from algae living in hot springs

DNA is heated to 95 deg C -> single chains

Solution is then cooled to 55 deg C; at 55 deg primers bind to ends of the single strand DNA

The temperature is now raised to 72 deg and the DNA polymerase causes the synthesis of new complementary strands to all the single strands

This is the end of the first cycle

The temperature is cycled from 95 to 55 to 72 over and over

The DNA is doubled at each cycle and at the end of 32 cycles it has been amplified 1 billion times

A cycle can be done in as little as 17 seconds, so it is possible to get a billion-fold amplification in an hour or less including set-up time

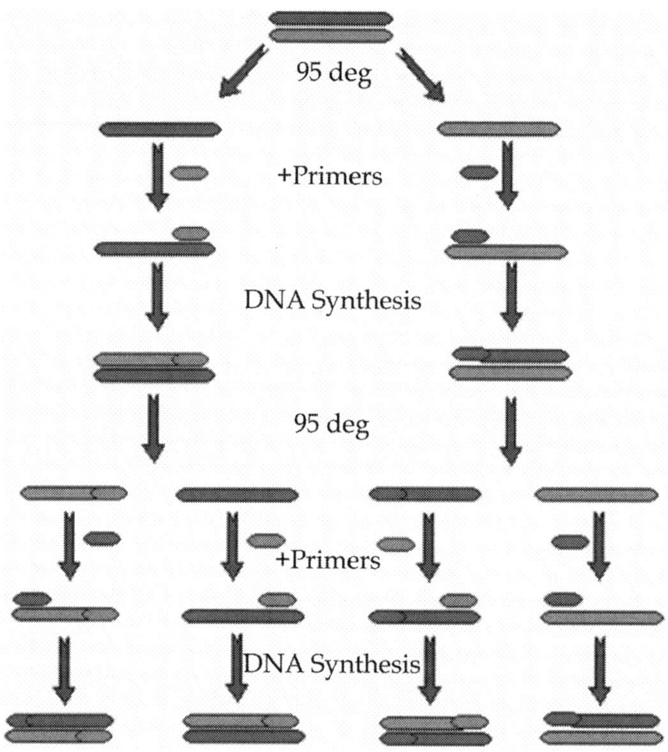

The figures shows 2 cycles of PCR

PCR can be used to dupicate single DNA molecules

It has been used to amplify DNA samples from extinct species

Cloned DNA in Colonies Can be Identified by Treating a Replica with Complementary DNA Probes

Often it is desirable to find bacterial colonies that have specific cloned DNA fragments, a specific gene for example

If some of the DNA sequence is known a short single strand radioactive probe can be made to search for the desired sequence

Most DNA probes have 15-50 bases

A mirror-image replica of the agar plate with the bacterial colonies is made

A piece of nitrocellulose paper is placed over the agar plate and pressed down

Some of the bacteria are transferred to the nitrocellulose

The replica is treated with alkali to break up the cells and is exposed to the probe

The probe binds to DNA having a complementary sequence, labelling those colonies that have it

Once the colonies having the desired gene have been identified the investigator can go back to the original plate and get living samples to culture

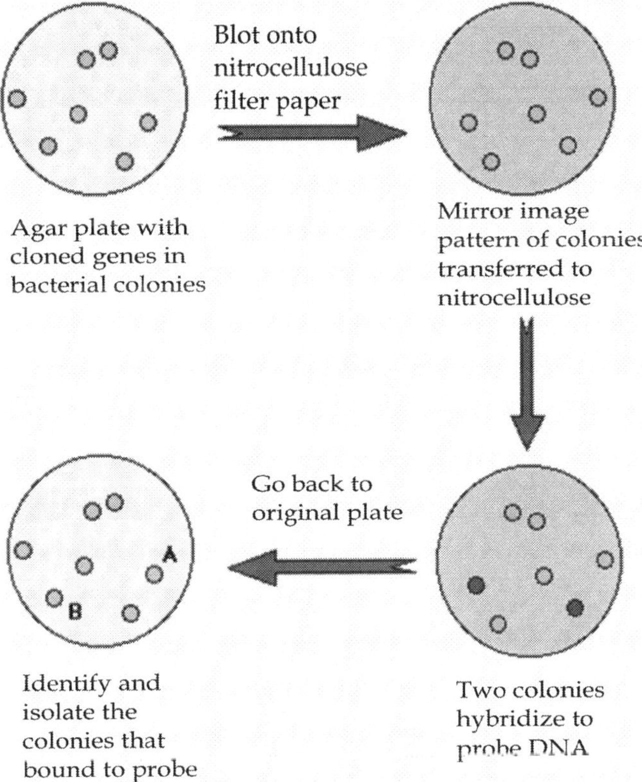

If a Gene is Expressed Antibodies Can be Used to Identify Proteins

Under certain conditions bacteria can express the protein coded by a cloned gene

Bacteria cannot express genomic DNA because it has introns

Can express complementary DNA, because it is made from messenger RNA (introns removed)

Genomic DNA can be expressed in eukaryotic cells, such as yeast, because they know how to handle introns

If a protein has been isolated antibodies can be made

The antibodies can be used to probe a replica plate to find colonies making the protein

This allows the gene for the protein to be isolated

Blotting Techniques Help to Analyze Restriction Fragments

Electrophoresis gels can also be analyzed with DNA probes by a Southern blot

After gel is run a replica is made by blotting the DNA onto a nitrocellulose or nylon filter

The replica is then treated with probes that reveal specific bands

Example of a Southern blot:

Suppose you are interested in a gene which has 3 restriction sites for a certain enzyme (marked by X in the picture, sample A); one of the sites occassionally mutates and disappears (sample B, from another person)

You have a probe which binds at the spot shown

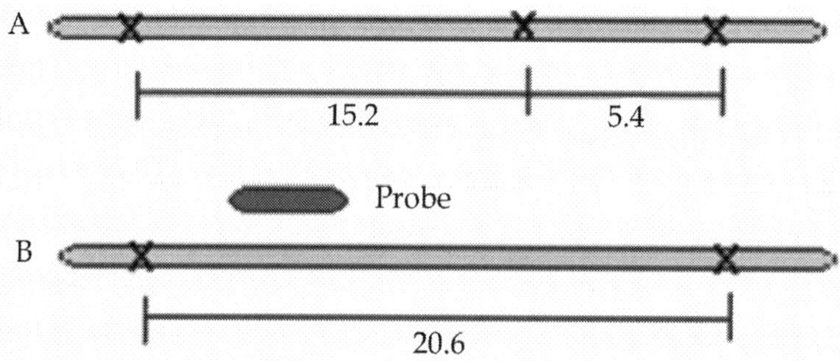

You extract the DNA from your cells and cut it with the restriction enzyme

This produces hundreds of restriction fragments because the enzyme cuts sites in other genes as well as the one you are interested in

If you do a Southern blot and then stain the DNA with a general stain such as ethidium bromide you will see a smear like this because the hundreds of bands run together

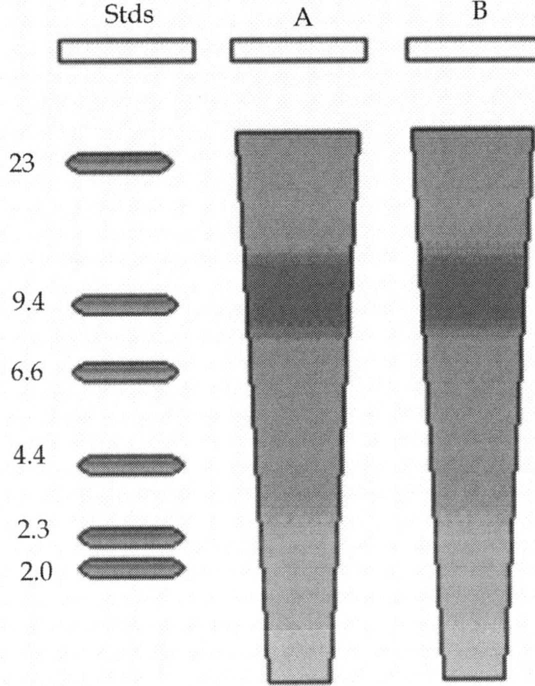

If you treat the replica with the probe only the bands that bind the probe will light up

The banding shows that when the mutation occurs the 15.2 kilobase band (sample A) disappears and is replaced by a 20.6 kilobase band (sample B)

Why doesn't the 5.4 kilobase band show in sample A? Look at the diagram above to see where the probe binds.

The Southern blot is named for the man who developed the technique

Other types of blots have been developed:

Northern blot: probes RNA

Western blot: probes protein, using antibodies

RFLPs: Different Individuals May Have Different Sizes of Restriction Fragments

RFLP = restriction fragment length polymorphism

A fancy way of saying, when you cut the genes up (with restriction enzymes) different people have different sized chunks

The A & B samples that we discussed Southern blotting are examples: the same enzyme gives a 20.6 kb fragment in one case, and a 15.2 plus a 5.4 kb fragment in the other case

To illustrate RFLPs further we will discuss 2 real cases, both associated with the sickle cell anemia mutation

Case 1: HpaI sites flanking the globin gene

If you take the DNA from 2 different individuals and compare the sequences in regions that do not code for genes you will find differences in 0.2 to 1% of the bases

Most of these differences are neutral since they do not affect coding for genes; they tend to accumulate because they are not eliminated by natural selection

Sometimes these variations in bases can be used as markers for genes

The variation must affect a restriction site

It must be very close to a gene

In 1978 Kan & Dozy (UC San Francisco) reported that restriction sites of the enzyme HpaI were associated with sickle cell anemia; they hoped the polymorphism could be used for diagnosis

The HpaI sites are in non-coding regions flanking the globin gene

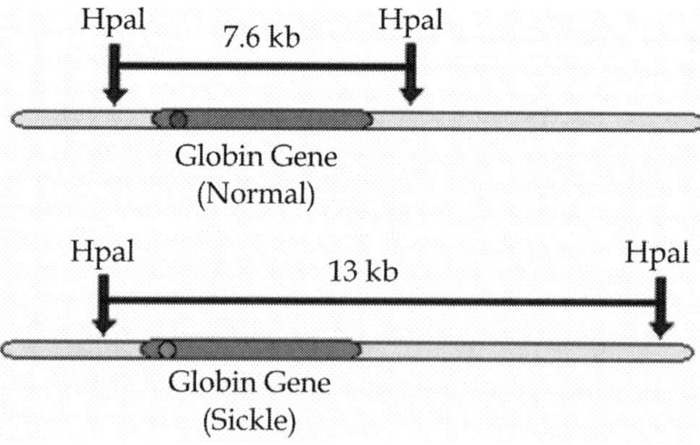

It was found that the HpaI site downstream from the gene was different in people with normal hemoglobin A and those with hemoglobin S

American blacks (about 60%) with hemoglobin S had a 13 kb restriction fragment

Those with normal hemoglobin A had a 7.6 kb restriction fragment

Probable origin of the association between the 2 genes:

Step 1: a mutation occurs producing a 13 kb restriction fragment in people with normal hemoglobin (probably in the Upper Volta region of Africa)

Step 2: the sickle mutation occurs in a person with the 13 kb restriction fragment and spreads throughout West Africa & Mediteranean

In other parts of the world the sickle mutation apparently occured in a person or several persons with the 7.6 kb restriction fragment (India, Saudi Arabia, East Africa)

This RFLP is of limited value for diagnosis because in some parts of the world sickle cell anemia is associated with the 13 kb fragment, while in others it is associated with the 7.6 kb fragment

Case 2: MstII site in the region with the sickle point mutation

The codon region near the sickle mutation is cut by several restriction enzymes

The most useful of these enzymes is MstII, which cuts in 3 places near the mutation site:

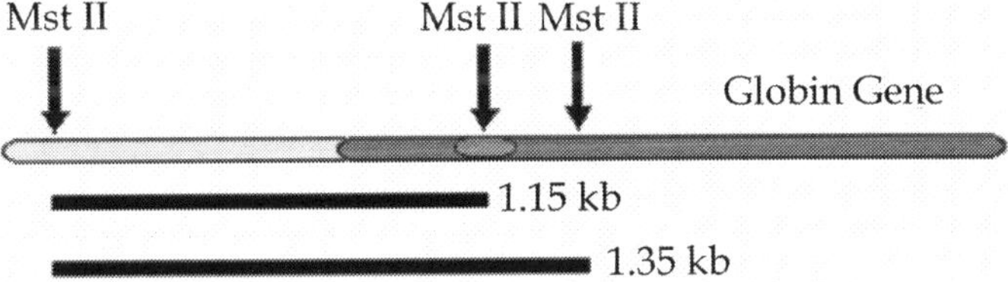

The sickle mutation destroys the middle MstII site

N in the MstII site code stands for any nucleotide (it can be G, for instance)

Hemoglobin A has an MstII site in codons 5 to 7, but changing an A to a T destroys the site in hemoglobin S

5 6 7	Codon
CCT-GAG-GAG	Hemoglobin A
CCT-GTG-GAG	Hemoglobin S
CCT-NAG-G	Mstll Site

In normal hemoglobin MstII will produce a fragment of 1.15 kb in this region

When the site is mutated the fragment enlarges to 1.35 kb

If you do a Southern blot test, this is the pattern you will see for people with the different types of hemoglobin:

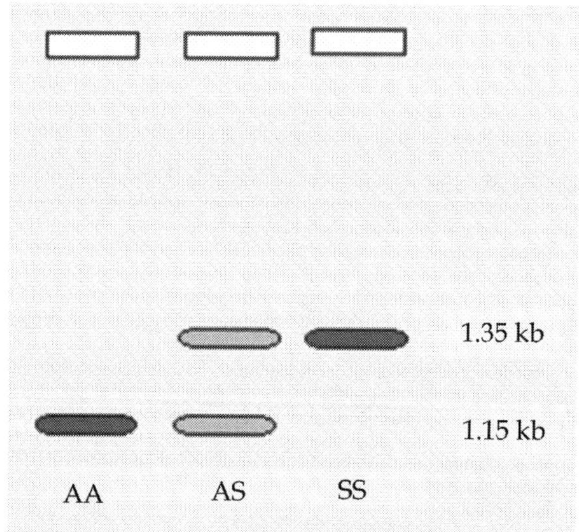

The AS individual has one gene of each type and this gives him 2 bands

Since the restriction site and mutation site overlap this test will give the correct diagnosis 100% of the time (excluding experimental error)

Gene Therapy Tries to Replace Damaged Genes

Cloning techniqes could be used to replace damaged genes with good ones

Genes have already been inserted into animal eggs to make transgenic study animals

Sickle cell anemia gene has been put into mice, for example

Gene insertion into the germline (eggs):

Can be used to prevent appearance of a disease

Fertilized egg removed from animal

DNA inserted into nucleus with micropipette

Egg put back into female for development

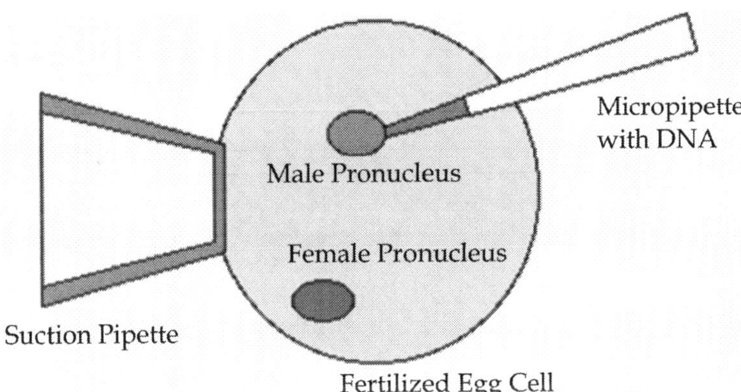

Technical problems

Gene must insert into egg DNA; insertion is random and often fails

Gene needs control elements such as promoters- these may be missing in random insertion

Attempts to target genes

Ethical problems

Many object to directly controlling heredity in this way

Gene inserted into adult animal

This is an attempt to cure a disease that is already present

Usually gene is put into a vector, such as a harmless virus, that can carry the gene into cells

Very difficult because millions of cells must receive the gene

CHAPTER - 23

Introduction to Plant Physiology

WHAT IS PLANT?

A. Definition - by most definitions, a plant:

- is multi-cellular;
- is non-motile
- has eukaryotic cells
- has cell walls comprised of cellulose
- is autotrophic; and
- exhibits alternation of generations - has a distinctive diploid (sporophyte) and haploid (gametophyte) phase.

B. Examples - the Plant Kingdom includes the angiosperms (flowering plants), gymnosperms (cone-bearing plants), ferns, and bryophytes (mosses & liverworts). Recent classification systems suggest that these organisms, in addition to the red algae and green algae, should be classified in the Plant Kingdom (Plantae).

Characteristics of Plants

A. Plants are dendritic

In other words, the basic shape of the plant body is dendritic - which means "tree-like" or "filamentous". The advantage of this shape is that it provides a large surface-to-volume (s/v) ratio which enables a plant to exploit a large area of the environment. In contrast, animals are more compact (spherical) to minimize their s/v ratio. Among other things, this is an advantage for motility. Surface-to-volume ratios are very important in many areas of biology.

B. Plants have indeterminate growth

Process by which a plant continues to grow and get larger throughout its life cycle. The advantage of this is that it allows the plant, especially roots, to grow into new areas. In contrast, determinate growth is where an organism or part reaches a certain size and then stops growing. This is characteristic of animals and some plant parts (*e.g.*, leaves, fruits).

C. Plants have an architectural design

In other words, the plant body is constructed like a building - modular (Silverton & Gordon). It is built of a limited number of units, each of which is relatively independent of the others and that are united into a single structure. Thus, just like a building is made of rooms, the leaves, stems and roots of a plant are analogous to a rooms in the building. Each room is somewhat independent, yet they all function together to make an integrated whole. You can seal off a room in a building, or remove a leaf or fruit, with little harm to the overall integrity of the structure. This is critical for plants to be able to add or remove parts (leaves, stems, flowers, fruits) as necessary. One conclusion is that because of their indeterminate growth and architectural design, ***plants are not limited by size.*** This gives plants the ability to colonize and exploit new areas for resources.

In contrast, an animal has a mechanical design. In other words, animals are built more like a machine, made of numerous, different parts that function together. The parts are highly integrated. Parts cannot be added or removed without reducing the efficiency of the operation of the whole. Animals are limited by size.

As a consequence, ***plants are not a static shape*** - plants constantly change shape by adding/loosing parts - by accumulating modular units. Animals don't change shape - they remain the same general shape throughout their life. Thus, growth in plants occurs by the addition of new "units" not enlargement.

D. Plants have a well developed ability to reproduce asexually

This can be viewed as a quick and energetically inexpensive way to expand the influence of the parent into a new location. One testable prediction from this hypothesis is that plants under nutrient stress should increase their rate of asexual reproduction.

E. Plants (may) exhibit heterophylly

Heterophylly refers to leaves with different shapes. For example, the aerial leaves of aquatic plants are entire but the submerged leaves are dissected. Sun leaves tend to be smaller and thicker than shade leaves. Dandelions are toothier when grown in a carbon dioxide enriched (700 vs. 350 ppm) environment. The leaves of the vine *Monstera* are more or less heart-shaped and pressed to the trunk as the vine climbs into the tropical forest canopy. Once in the canopy, the leaves take on the mature form with slits and holes.

F. Leaves are arranged to minimize overlapping

Phyllotaxy is the fancy term for leaf arrangement. Interestingly, phyllotaxy patterns have always been shown to be related to Fibonacci number series (1, 1, 2, 3, 5, 8, 13, 21....etc).

G. Plants can forage

The growth patterns of plants, especially vines and plants with stolons (runners), are similar to the foraging tactics of animals. As an example, rhizomes of *Hydrocotyle* veer from patches of grass to avoid competition. A brief overview of the anatomy of a clonal plant like *Glechoma hederacea* (ground ivy): parent plant, stolon (internode), ramet (individual of a clone).

Thus, plant growth is essentially analogous to animal behavior. One of the first to express this idea was Arber (1950; *The Natural Philosophy of Plant Form*. Cambridge). She said, "Among plants, form may be held to include something corresponding to behavior in the zoological field...for most though but not for all plants the only available forms of action are either growth, or ascending of parts, both of which involve a change in the size and form of the organism."

What is Plant Physiology?

A. Definitions (numerous) - Plant physiology is the study of:

- the functions and processes occurring in plants
- the vital processes occurring in plants
- how plants work

B. In essence, plant physiology is a study of the plant way of life, which include various aspects of the plant lifestyle and survival including: metabolism, water relations, mineral nutrition, development, movement, irritability (response to the environment), organization, growth, and transport processes.

Plant physiology is a branch of plant sciences that aims to understand how plants live and function. Its ultimate objective is to explain all life processes of plants by a minimal number of comprehensive principles founded in chemistry, physics, and mathematics.

Fundamental processes such as photosynthesis, respiration, plant nutrition, plant hormone functions, tropisms, nastic movements, photoperiodism, photomorphogenesis, circadian rhythms, environmental stress physiology, seed germination, dormancy and stomata function and transpiration, both part of plant water relations, are studied by plant physiologists.

Plant physiology seeks to understand all the aspects and manifestations of plant life. In agreement with the major characteristics of organisms, it is usually divided into three major parts:

(1) the physiology of nutrition and metabolism, which deals with the uptake, transformations, and release of materials, and also their movement within and between the cells and organs of the plant;

(2) the physiology of growth, development, and reproduction, which is concerned with these aspects of plant function; and

(3) environmental physiology, which seeks to understand the manifold responses of plants to the environment. The part of environmental physiology which deals with effects of and adaptations to adverse conditions—and which is receiving increasing attention—is called stress physiology.

History

Sir Francis Bacon published one of the first plant physiology experiments in 1627 in the book, *Sylva Sylvarum*. Bacon grew several terrestrial plants, including a rose, in water and concluded that soil was only needed to keep the plant upright. Jan Baptist van Helmont published what is considered the first quantitative experiment in plant physiology in 1648. He grew a willow tree for five years in a pot containing 200 pounds of oven-dry soil. The soil lost just two ounces of dry weight and van Helmont concluded that plants get all their weight from water, not soil. In 1699, John Woodward published experiments on growth of spearmint in different sources of water. He found that plants grew much better in water with soil added than in distilled water.

Stephen Hales is considered the Father of Plant Physiology for the many experiments in the 1727 book; though Julius von Sachs unified the pieces of plant physiology and put them together as a discipline. His *Lehrbuch der Botanik* was the plant physiology bible of its time.

Researchers discovered in the 1800s that plants absorb essential mineral nutrients as inorganic ions in water. In natural conditions, soil acts as a mineral nutrient reservoir but the soil itself is not essential to plant growth. When the mineral nutrients in the soil are dissolved in water, plant roots absorb nutrients readily, soil is no longer required for the plant to thrive. This observation is the basis for hydroponics, the growing of plants in a water solution rather than soil, which has become a standard technique in biological research, teaching lab exercises, crop production and as a hobby.

Scope

The field of plant physiology includes the study of all the internal activities of plants—those chemical and physical processes associated with life as they occur in plants. This includes study at many levels of scale of size and time. At the smallest scale are molecular interactions of photosynthesis and internal diffusion of water, minerals, and nutrients. At the largest scale are the processes of plant development, seasonality, dormancy, and reproductive control. Major subdisciplines of plant physiology include phytochemistry (the study of the biochemistry of plants) and phytopathology (the study of disease in plants). The scope of plant physiology as a discipline may be divided into several major areas of research.

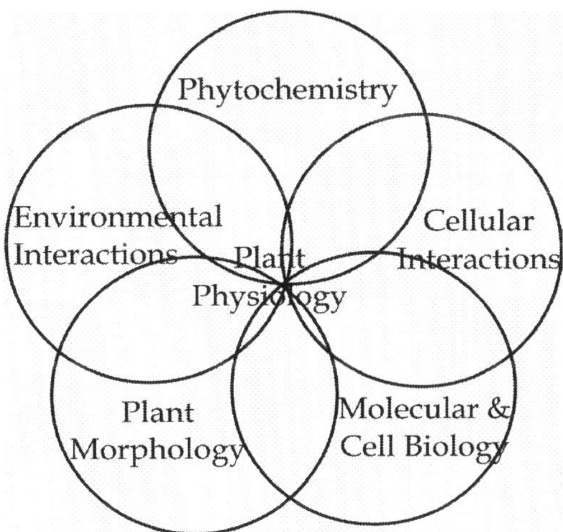

Five key areas of study within plant physiology

First, the study of phytochemistry (plant chemistry) is included within the domain of plant physiology. To function and survive, plants produce a wide array of chemical compounds not found in other organisms. Photosynthesis requires a large array of pigments, enzymes, and other compounds to function. Because they cannot move, plants must also defend themselves chemically from herbivores, pathogens and competition from other plants. They do this by producing toxins and foul-tasting or smelling chemicals. Other compounds defend plants against disease, permit survival during drought, and prepare plants for dormancy. While other compounds are used to attract pollinators or herbivores to spread ripe seeds.

Secondly, plant physiology includes the study of biological and chemical processes of individual plant cells. Plant cells have a number of features that distinguish them from cells of animals, and which lead to major differences in the way that plant life behaves and responds differently from animal life. For example, plant cells have a cell wall which restricts the shape of plant cells and thereby limits the flexibility and mobility of plants. Plant cells also contain chlorophyll, a chemical compound that interacts with light in a way that enables plants to manufacture their own nutrients rather than consuming other living things as animals do.

Thirdly, plant physiology deals with interactions between cells, tissues, and organs within a plant. Different cells and tissues are physically and chemically specialized to perform different functions. Roots and rhizoids function to anchor the plant and acquire minerals in the soil. Leaves catch light in order to manufacture nutrients. For both of these organs to remain living, minerals that the roots acquire must be transported to the leaves, and the nutrients manufactured in the leaves must be transported to the roots. Plants have developed a number of ways to achieve this transport, such as vascular tissue, and the functioning of the various modes of transport is studied by plant physiologists.

Fourthly, plant physiologists study the ways that plants control or regulate internal functions. Like animals, plants produce chemicals called hormones which are produced in one part of the plant to signal cells in another part of the plant to respond. Many flowering plants bloom at the appropriate time because of light-sensitive compounds that respond to the length of the night, a phenomenon known as photoperiodism. The ripening of fruit and loss of leaves in the winter are controlled in part by the production of the gas ethylene by the plant.

Finally, plant physiology includes the study of how plants respond to conditions and variation in the environment, a field known as environmental physiology. Stress from water loss, changes in air chemistry, or crowding by other plants can lead to changes in the way a plant functions. These changes may be affected by genetic, chemical, and physical factors.

Biological Pigments

Among the most important molecules for plant function are the pigments. Plant pigments include a variety of different kinds of molecules, including porphyrins, carotenoids, and anthocyanins. All biological pigments selectively absorb certain wavelengths of light while reflecting others. The light that is absorbed may be used by the plant to power chemical reactions, while the reflected wavelengths of light determine the color the pigment appears to the eye.

Chlorophyll is the primary pigment in plants; it is a porphyrin that absorbs red and blue wavelengths of light while reflecting green. It is the presence and relative abundance of chlorophyll that gives plants their green color. All land plants and green algae possess two forms of this pigment: chlorophyll *a* and chlorophyll *b*. Kelps, diatoms, and other photosynthetic heterokonts contain chlorophyll *c* instead of *b*, while red algae possess only chlorophyll *a*. All chlorophylls serve as the primary means plants use to intercept light to fuel photosynthesis.

Carotenoids are red, orange, or yellow tetraterpenoids. They function as accessory pigments in plants, helping to fuel photosynthesis by gathering wavelengths of light not readily absorbed by chlorophyll. The most familiar carotenoids are carotene (an orange pigment found in carrots), lutein (a yellow pigment found in fruits and vegetables), and lycopene (the red pigment responsible for the color of tomatoes). Carotenoids have been shown to act as antioxidants and to promote healthy eyesight in humans.

Anthocyanins (literally "flower blue") are water-soluble flavonoid pigments that appear red to blue, according to pH. They occur in all tissues of higher plants, providing color in leaves, stems, roots, flowers, and fruits, though not always in sufficient quantities to be noticeable. Anthocyanins are most visible in the petals of flowers, where they may make up as much as 30% of the dry weight of the tissue. They are also responsible for the purple color seen on the underside of tropical shade plants such as *Tradescantia zebrina*. In these plants, the anthocyanin catches light that has passed through the leaf and reflects it back towards regions bearing chlorophyll, in order to maximize the use of available light

Betalains are red or yellow pigments. Like anthocyanins they are water-soluble, but unlike anthocyanins they are indole-derived compounds synthesized from tyrosine. This class of pigments is found only in the Caryophyllales (including cactus and amaranth), and never co-occur in plants with anthocyanins. Betalains are responsible for the deep red color of beets, and are used commercially as food-coloring agents. Plant physiologists are uncertain of the function that betalains have in plants which possess them, but there is some preliminary evidence that they may have fungicidal properties.

Signals and regulators

Plants produce hormones and other growth regulators which act to signal a physiological response in their tissues. They also produce compounds such as phytochrome that are sensitive to light and which serve to trigger growth or development in response to environmental signals.

Plant hormones

Plant hormones, known as plant growth regulators (PGRs) or phytohormones, are chemicals that regulate a plant's growth. According to a standard animal definition, hormones are signal molecules produced at specific locations, that occur in very low concentrations, and cause altered processes in target cells at other locations. Unlike animals, plants lack specific hormone-producing tissues or organs. Plant hormones are often not transported to other parts of the plant and production is not limited to specific locations.

Plant hormones are chemicals that in small amounts promote and influence the growth, development and differentiation of cells and tissues. Hormones are vital to plant growth; affecting processes in plants from flowering to seed development, dormancy, and germination. They regulate which tissues grow upwards and which grow downwards, leaf formation and stem growth, fruit development and ripening, as well as leaf abscission and even plant death.

The most important plant hormones are abscissic acid (ABA), auxins, ethylene, gibberellins, and cytokinins, though there are many other substances that serve to regulate plant physiology.

Photomorphogenesis

While most people know that light is important for photosynthesis in plants, few realize that plant sensitivity to light plays a role in the control of plant structural development (morphogenesis). The use of light to control structural development is called photomorphogenesis, and is dependent upon the presence of specialized photoreceptors, which are chemical pigments capable of absorbing specific wavelengths of light.

Plants use four kinds of photoreceptors: phytochrome, cryptochrome, a UV-B photoreceptor, and protochlorophyllide *a*. The first two of these, phytochrome and cryptochrome, are photoreceptor proteins, complex molecular structures formed by joining a protein with a light-sensitive pigment. Cryptochrome is also known as the UV-A photoreceptor, because it absorbs ultraviolet light in the long wave "A" region. The UV-B receptor is one or more compounds not yet identified with certainty, though some evidence suggests carotene or riboflavin as candidates. Protochlorophyllide *a*, as its name suggests, is a chemical precursor of chlorophyll.

The most studied of the photoreceptors in plants is phytochrome. It is sensitive to light in the red and far-red region of the visible spectrum. Many flowering plants use it to regulate the time of flowering based on the length of day and night (photoperiodism) and to set circadian rhythms. It also regulates other responses including the germination of seeds, elongation of seedlings, the size, shape and number of leaves, the synthesis of chlorophyll, and the straightening of the epicotyl or hypocotyl hook of dicot seedlings.

Photoperiodism

Many flowering plants use the pigment phytochrome to sense seasonal changes in day length, which they take as signals to flower. This sensitivity to day length is termed photoperiodism. Broadly speaking, flowering plants can be classified as long day plants, short day plants, or day neutral plants, depending on their particular response to changes in day length. Long day plants require a certain minimum length of daylight to initiate flowering, so these plants flower in the spring or summer. Conversely, short day plants flower when the length of daylight falls below a certain critical level. Day neutral plants do not initiate flowering based on photoperiodism, though some may use temperature sensitivity (vernalization) instead.

Although a short day plant cannot flower during the long days of summer, it is not actually the period of light exposure that limits flowering. Rather, a short day plant requires a minimal length of uninterrupted darkness in each 24 hour period (a short daylength) before floral development can begin. It has been determined experimentally that a short day plant (long night) does not flower if a flash of phytochrome activating light is used on the plant during the night.

Plants make use of the phytochrome system to sense day length or photoperiod. This fact is utilized by florists and greenhouse gardeners to control and even induce flowering out of season, such as the *Poinsettia*.

Environmental physiology

Paradoxically, the subdiscipline of environmental physiology is on the one hand a recent field of study in plant ecology and on the other hand one of the oldest. Environmental physiology is the preferred name of the subdiscipline among plant physiologists, but it goes by a number of other names in the applied sciences. It is roughly synonymous with ecophysiology, crop ecology, horticulture and agronomy. The particular name applied to the subdiscipline is specific to the viewpoint and goals of research. Whatever name is applied, it deals with the ways in which plants respond to their environment and so overlaps with the field of ecology.

Environmental physiologists examine plant response to physical factors such as radiation (including light and ultraviolet radiation), temperature, fire, and wind. Of particular importance are water relations (which can be measured with the Pressure bomb) and the stress of drought or inundation, exchange of gases with the atmosphere, as well as the cycling of nutrients such as nitrogen and carbon.

Environmental physiologists also examine plant response to biological factors. This includes not only negative interactions, such as competition, herbivory, disease and parasitism, but also positive interactions, such as mutualism and pollination.

Tropisms and nastic movements

Plants may respond both to directional and nondirectional stimuli. A response to a directional stimulus, such as gravity or sunlight, is called a tropism. A response to a nondirectional stimulus, such as temperature or humidity, is a nastic movement.

Tropisms in plants are the result of differential cell growth, in which the cells on one side of the plant elongate more than those on the other side, causing the part to bend toward the side with less growth. Among the common tropisms seen in plants is phototropism, the bending of the plant toward a source of light. Phototropism allows the plant to maximize light exposure in plants which require additional light for photosynthesis, or to minimize it in plants subjected to intense light and heat. Geotropism allows the roots of a plant to determine the direction of gravity and grow downwards. Tropisms generally result from an interaction between the environment and production of one or more plant hormones.

In contrast to tropisms, nastic movements result from changes in turgor pressure within plant tissues, and may occur rapidly. A familiar example is thigmonasty (response to touch) in the Venus fly trap, a carnivorous plant. The traps consist of modified leaf blades which bear sensitive trigger hairs. When the hairs are touched by an insect or other animal, the leaf folds shut. This mechanism allows the plant to trap and digest small insects for additional nutrients. Although the trap is rapidly shut by changes in internal cell pressures, the leaf must grow slowly to reset for a second opportunity to trap insects.

CHAPTER - 24

Water, Diffusion and Osmosis

Water, of its very nature, as it occurs automatically in the process of cosmic evolution, is fit, with a fitness no less marvelous and varied than that fitness of the organism which has been won by the process of adaptation in the course of organic evolution.

WATER IS ABSOLUTELY ESSENTIAL FOR ALL LIVING ORGANISMS

The evidence:

1. Most organisms are comprised of at least 70% or more water. Some plants, like a head of lettuce, are made up of nearly 95% water;
2. When organisms go dormant, they loose most of their water. For example, seeds and buds are typically less than 10% water, as are desiccated rotifers, nematodes and yeast cells;
3. Earth is the water planet (that's why astronomers get so excited about finding water in space).
4. Water is the limiting resource for crop productivity in most agricultural systems.

WATER IS IMPORTANT BECAUSE IT IS POLAR AND READILY FORMS HYDROGEN BONDS

A. Water is Polar

In other words, the water molecule has a positively-charged (hydrogen side) and negatively-charged side (oxygen). This occurs because:

1. the hydrogen atoms are arranged at an angle of about 105 degrees;
2. the covalent bond between O-H is polarized. This is caused by an unequal sharing of electrons between these atoms which, in turn, results in a slight negative charge on the oxygen atom (electronegative) and slight positive charge on the hydrogen;
3. oxygen has an unshared pair of electrons (the molecule is tetrahedral-shaped).

B. Hydrogen Bonds

The 'fancy' definition of a hydrogen bond is that it is a weak bond that forms between a hydrogen atom that is covalently bonded to an electronegative atom (like oxygen) and another electronegative atom. In other words, a positively-charged hydrogen atom is attracted to a negatively-charged oxygen.

The end result is that water readily forms hydrogen bonds with itself and other polar molecules. When likes attract it is termed **cohesion** (i.e., hydrogen bonds between water molecules). When unlikes attract, it is called **adhesion** (i.e., when a paper towel absorbs water, water and cellulose adhere to one another). Cohesion and adhesion are responsible for **capillary action**, the movement of water up a thin tube.

In liquid water, hydrogen bonds between water molecules are continuously made and broken. The molecules can even form temporary "quasi-crystalline" areas. Individually, each hydrogen bond is weak (20 kJ mol^{-1}), but collectively they give water many unique properties (a Marxist molecule!).

The Properties of Water

A. Water is a liquid at physiological temperatures (i.e., between 0-100^{0}C)

In other words, water has a **high boiling point** and a **high melting point** when compared to other similar-sized molecules such as ammonia, carbon dioxide, hydrogen sulfide. These other molecules are gases at room temperature. This is important because if life exists anywhere, we predict that it occurs between approx. 0 and 100 C. Temperatures much below 0 are too cold to permit significant chemistry for metabolism; temperatures above 100 tend to disrupt bonds.

B. Water has a high heat of vaporization

In other words, it takes a lot of energy (ca. 44 kJ mol^{-1}) to convert water from a liquid to a gas; or stated another way, *Water resists evaporation.* This property is responsible for water's use in **evaporative cooling** systems, hence the reason dog's pant, people perspire, and leaves transpire.

C. Water has a high specific heat (heat capacity)

It takes a lot of energy (4.184 J g^{-1} C^{-1}, or the non-SI unit is the calorie where 1 cal = 4.184 J) to raise the temperature of water (because it requires a lot of energy to break/make hydrogen bonds). Thus, water is slow to heat up and cool down, or stated another way, *water resists temperature changes.* This is why you can swim reasonably comfortably in the Sag in late fall but not the spring. In contrast, a sidewalk has low specific heat - it heats up quickly (try walking barefoot on a sidewalk in summer on a sunny day), but cools down quickly. This property is important in water's role as a **thermal buffer**. It's not surprising that desert plants are succulent - to help resist temperature fluctuations.

D. Water has a high heat of fusion

It takes a lot of energy to convert water from a solid to a liquid, or put another way, *water resists freezing*. Energy is required to break the collective hydrogen bonds holding water in its solid configuration. Conversely, a lot of energy (6 kJ mol^{-1}) must be released by water to freeze. This property is used by citrus growers - prior to a light freeze they spray fruit with water; ice forms releasing the heat of fusion which will help protect the crop from serious damage.

E. Water has a high surface tension

It takes a lot of energy to break through the surface of water, because water molecules at the surface are attracted (cohesion) to others within the liquid much more than they are to air. Thus, *water acts as though it has a skin.* This phenomenon is important at air/water interfaces and explains why: (1) water rises up a thin tube (capillary action); (2) raindrops are round (the molecules at the surface attract one another); (3) water striders and other bugs can "walk on water"; (4) a meniscus forms; and (5) a belly-whopper into a pool of ammonia would not hurt nearly as much as one into water.

F. The density of water decreases on crystallization

Good thing too, or else ice fishing would be a moist business. This occurs because when ice forms each water molecule is hydrogen bonded to exactly four others. At four degrees, water is it's densest, and each water molecule is attracted to slightly more than four others. Thus, as water cools it gets denser and denser until it reaches 4 C, then, it gets less dense. And ice floats.

G. Water is a universal solvent

Water dissolves more different kinds of molecules than any other solvent. Hydrophilic (water-loving) molecules dissolve readily in water (likes dissolve likes), hydrophobic (water-fearing) ones do not.

H. Water has high tensile strength and incompressibility

In other words, if you put water in a tube and put a piston on either end, you won't be able to push the pistons together. Thus, water is good for hydraulic systems because when it is squeezed it doesn't compress and produces positive pressures (hydrostatic pressures). This pressure provides the driving force for cell growth and other plant movements. The pressure is measured in units of Pascals (or actually MegaPascals, MPa). One MPa is approximately equal to ten atmospheres or 10 bars.

In a similar fashion, if you fill the tube with water, remove any air bubbles, and then pull the pistons away from one another, the water column resists breaking. This will result in a suction on the water column - just like putting your finger on the end of a syringe and pulling back the plunger. Negative pressures (**tensions**) can develop in the water column. Very sizable tensions can be generated in a thin water column. However, **cavitation,** when air comes out of solution at negative pressures, can be a problem.

I. Water is transparent to light

This is important because chloroplasts (inside a cell) are obviously surrounded by water. If water were opaque, plants couldn't photosynthesize. From an ecological perspective, the penetration of in water determines the distribution of aquatic plants.

J. Water is chemically inert

It doesn't react unless it is enzymatically designed to do so.

K. Water dissociates into protons and hydroxide ions

This serves as the basis for the pH system.

L. Water affects the shape, stability & properties of biological molecules.

For example, many ions (such as sodium) and molecules (such as DNA and wall components) are normally hydrated. This means that water is hydrogen bonded to them and in some cases (i.e., sodium) forms a hydration shell around them.

Functions of Water

In addition to the functions mentioned above, water:

1. is a major component of cells
2. is a solvent for the uptake and transport of materials
3. is a good medium for biochemical reactions
4. is a reactant in many biochemical reactions (*i.e.*, photosynthesis)
5. provides structural support via turgor pressure (*i.e.*, leaves)
6. is the medium for the transfer of plant gametes (sperms swim to eggs in water, some aquatic plants shed pollen underwater)
7. offspring (propagule) dispersal (think "coconut")
8. plant movements are the result of water moving into and out of those parts (*i.e.*, diurnal movements, stomatal opening, flower opening)
9. cell elongation and growth
10. thermal buffer
11. perhaps most importantly, water has directed the evolution of all organisms. You can think of morphological features of organisms as a consequence of water availability. For example, consider organisms growing in xeric (dry), mesic (moderate) and hydric (aquatic) environments.

Acids and Bases

Water ionizes to a small degree to form a hydrogen ion (or proton) and hydroxide ion (OH-). In reality, two water molecules form a hydronium ion (H_30^+) and a hydroxide ion (OH^-).

In pure water,

$[H^+]$ = $[OH^-]$ This solution is neutral

$[H^+]$ > $[OH^-]$ Then, the solution is an acid (acidic)

$[H^+]$ < $[OH^-]$ Then the solution is a base (alkaline)

Thus:

1. An acid is a substance that increases the $[H^+]$, or as the chemists say, is a proton donor.

   eg. $HCl \rightarrow H^+ + Cl^-$
2. A base is a substance that increases the $[OH^-]$; or from the perspective of a proton, a base is a substance that decreases the proton concentration; it is a proton acceptor.

e.g. $NaOH \rightarrow Na^+ + OH^-$ (accepts protons to make water)

e.g. NH_3 (ammonia) + $H^+ \rightarrow NH_4^+$ (ammonium ion)

The pH Scale:

pH is the scale to express the degree of acidity (or alkalinity) of a solution. The scale ranges from 0 to 14 where 1 is highly acidic, 7 is neutral, and 14 is highly alkaline.

As the pH increases, the $[H^+]$ decreases and the $[OH^-]$ increases

As the pH decreases, the $[H^+]$ increases and the $[OH^-]$ decrease

pH = - log[H+]

Points to note:

1. the pH scale is based on proton concentration; and
2. the pH scale is logarithmic, there is a 10-fold difference in concentration between each pH unit.

at pH 7:

$$[H^+] = [OH^-]$$

$$0.0000001 \text{ mol } H^+ \text{ liter}^{-1} = 10^{-7} H^+ = 0.0000001 \text{ } OH^- \text{ liter}^{-1}$$

$$pH = \log [H^+]$$

$$= -\log[10^{-7}]$$

$$= -(-7)$$

$$= 7$$

The products of $[H^+] \times [OH^-]$ always equals 10^{-14}. Thus, you can always determine concentration of one if you know the other. For example, if the $[H^+] = 10^{-2}$, then the $[OH^-]$ is 10^{-12}.

Living systems are very sensitive to pH

Organisms must maintain pH within tolerable ranges. This is a good example of homeostasis.

A buffer is a solution that resists fluctuations in pH when additional OH- or H+ are added. They maintain a constant pH and usually consist of a proton donor and a proton acceptor. [e.g., blood pH must be between 7.36 (venous) and 7.41 (arterial). The carbonic acid/bicarbonate buffer helps to maintain pH:

H_2CO_3 (carbonic acid; proton donor) ß ׀ $H^+ + HCO_3^-$ (bicarbonate ion; proton acceptor)

Water Movement

There are two major ways to move molecules:

A. Bulk (or Mass) Flow

This is the mass movement of molecules in response to a pressure gradient. The molecules move from hi à low pressure, following a pressure gradient. A good example would be a faucet. When you turn a faucet on, water comes out. This occurs because the water in the tap is under pressure relative to the air outside the faucet. A toilet is another example; high pressure in the tank/bowl but lower pressure in the sewer system. Some molecular movements rely on bulk flow which requires a mechanism to generate the pressure gradient. For example, animals have evolved a pump (*i.e.,* heart) that is designed for the bulk flow of molecules through the circulatory system.

B. Diffusion

The net, random movement of individual molecules from one area to another. The molecules move from [hi] α [low], following a concentration gradient. Another way of stating this is that the molecules move from an area of high free energy (higher concentration) to one of low free energy (lower concentration). The net movement stops when a **dynamic equilibrium** is achieved.

Imagine opening a bottle of perfume containing volatile essential oils in a very, very still room. Initially, the essential oils are concentrated in a corner of the room. As the molecules move randomly, in every different direction, over time they will eventually appear throughout the room. Ultimately the essential oils will reach a point, dynamic equilibrium, at which they are evenly distributed throughout the room. At this point the molecules are still moving. They continue to move randomly in every direction. The only difference is that there is no net change in the overall distribution of the perfume in the room.

Now imagine that the room is divided by a partition with holes (which is analogous to a membrane). If we place a drop of perfume on one side of the partition and then count at intervals the number of essential oil molecules on either side of the partition and graph the results:

insert: plot # molecules vs. time on both sides of the partition.

We will observe that the number of molecules on one side will decrease while the other will increase until they reach dynamic equilibrium. At equilibrium the molecules continue to move randomly, back and forth from one side of the partition to the other. Hence the number of molecules on either side of the partition at any given time is simply chance. The number oscillates about the midpoint.

A caveat

Although this theoretical example can help us to better understand the nature of diffusion, *it is technically wrong*. The molecular movement attributed to diffusion in this example is really due to air movements in the room, or convection. True examples of diffusion are hard to come by (see Vogel, 1994; Wheatley, 1993). Nevertheless, it serves our purpose to illustrate the general concept of diffusion.

C. Osmosis

This is a specialized case of diffusion; it represents the diffusion of a solvent (typically water) across a membrane.

D. Dialysis

Another specialized case of diffusion; it is the diffusion of solute across a semi-permeable membrane. Example – consider a cell containing a sugar dissolved in water. If water (the solvent) moves out of the cell into the surroundings it moves osmotically; if the sugar (solute) moves into the surroundings, it is an example of dialysis.

Factors influencing the rate of diffusion –

Several factors influence the rate of diffusion. These include:

A. Concentration gradient- As previously stated, solutes move from an area of high concentration to one of lower concentration; in other words, in response to a concentration gradient. Although this is true for most solutes, it is NOT important for water. The concentration of water (55.2 - 55.5 mol L^{-1}) is nearly constant under all conditions (*i.e.*, MW = 18 g/mol, and 1000 g/liter; thus, 1000/18 = 55.5 mol/L).

Fick's Law - is an equation that relates the rate of diffusion to the concentration gradient (C1 – C2) and resistance (r). Diffusion rate, also called flux density (J_s, in units of mol m^{-2} s^{-1}) can be expressed in the simplified version of Fick's equation as:

$$J_s = (C1 - C2) / r$$

The take-home-lessons from this equation are that the rate of diffusion are:

1. the rate of diffusion is directly proportional to the concentration gradient. The greater the difference in concentration between two areas, the greater the rate of diffusion. Thus, when the gradient is zero, there will be no net diffusion, diffusion will only occur so long as a concentration gradient exists;

2. the rate of diffusion is indirectly proportional to resistance. In other words, the greater the resistance to diffusion, the lower the rate of diffusion. Resistance refers to anything that reduces the rate of diffusion such as the partition in our perfume example. The width of the partitions is a resistance; the wider the partitions, the lower the resistance. And, the membrane is a resistance to the movement of ions and other charged substances in or out of cells; and

3. the rate of diffusion is inversely proportional to distance traveled (also a function of resistance). For example, some typical diffusion rates for water are 10 μm - 0.1 sec; 100 μm -1 sec; and 1 mm - 100 sec. As the text demonstrates nicely, diffusion is effective over short distances, but is pathetically slow over long distances.

B. Molecular Speed- According to kinetic theory, particles like atoms and molecules are in always in motion at temperatures above absolute zero (0 K = -273 C). The take-home-lesson is that molecular movement is:

1. directly proportional to temperature; and

2. indirectly related to molecular weight (heavier particles move more slowly than lighter, smaller ones). At room temperature, the average velocity of a molecule is fast - about 2 km/sec (=3997 mph!).

C. Temperature - increases the rate of molecular movement, therefore, increases the rate of diffusion

D. Pressure - increases speed of molecules, therefore, increase the rate of diffusion

E. Solute effect on the chemical potential of the solvent- Solute particles decrease the free energy of a solvent. The critical factor is the number of particles, not charge or particle size. Essentially solvent molecules, such as water in a biological system, move from a region of greater mole fraction to a region where it has a lower mole fraction. The mole fraction of solvent = # solvent molecules/ total (# solvent molecules + # solute molecules). This is particularly important in the movement of water. Water moves from an area of higher mole fraction or higher energy to an area of lower mole fraction or lower energy.

Measure of the energy state of water. This is a particularly important concept in plant physiology because it determines the direction and movement of water.

A. First, some definitions:

1. Free energy of water - energy available to do work (J = n m)

2. Chemical potential (μ) - free energy/unit quantity (usually per mole) ($J\ mol^{-1}$)

3. Water potential (ψ_w) - chemical potential of water, compared to pure water at the same temperature and pressure. The units are in pressure because: (a) plant cells under pressure (remember the wall?); and (b) it is easier to measure pressure.
4. Derivation of units - Water potential is official defined as the chemical potential of water divided by the partial molar volume. divide chemical potential (J mol^{-1}) by the partial molar volume of water (L mol^{-1}). This can be simplified in two ways:

   J mol^{-1}/ L mol^{-1} = J / liter = energy / volume = (weight x distance)/ area x distance = force/area = pressure

   or looking at it another way

   J mol^{-1}/ L mol^{-1} = n x m mol^{-1}/ m^3 mol^{-1} = n m^{-2} = MPa
5. Pressure is measured in MPa (megapascals). 1 MPa = 10 bars = 10 atm. (as an aside, 1 atm = 760 mm Hg = 14. 7 lbs sq in^{-1})

B. Equation for water potential (*must account for the factors that influence the diffusion of water and other substances*):

$$y_w = y_p + y_s + y_g$$

where y_w = water potential; y_p = pressure potential; ys = solute or osmotic potential; and y_g = gravity potential.

1. Solute (or osmotic) potential (y_s)

This is the contribution due to dissolved solutes. Solutes always decrease the free energy of water, thus there contribution is always negative. The solute potential of a solution can be calculated with the van't Hoff equation: $Ø_s$ = -miRT where m = molality (moles/1000 g); i = ionization constant (often 1.0); R = gas constant (0.0083 liter x MPa/mol deg); and T = temperature (K).

2. Pressure (or Pressure Potential; y_p)

Due to the pressure build up in cells thanks to the wall. It is usually positive, although may be negative (tension) as in the xylem. Pressure can be measured with an osmometer.

3. Matric potential

This is the contribution to water potential due to the force of attraction of water for colloidal, charged surfaces. It is negative because it reduces the ability of water to move. In large volumes of water it is very small and usually ignored. However, it can be very important in the soil, especially when referring to the root/soil interface.

4. Gravity (y_g)

Contributions due to gravity which is usually ignored unless referring to the tops of tall trees.

C. The water potential of pure water is zero. Water potentials in intact plant tissue are usually negative (because of the large quantities of dissolved solutes in cells).

D. Water potential is the sum of the contributions of the various factors that influence water potential

E. Measuring Water Potential - we will discuss the following techniques in class/lab:

1. Pressure bomb - a steel chamber that can be pressurized, usually with nitrogen. The sample is placed in the chamber with the petiole or surface exposed through a hole in the lid. The sample is pressurized and the pressure that is required to force water to appear on the cut surface is assumed to be equivalent to the water potential of the tissue.
2. Chardakov Method - Dye Drop method
3. Gravimetric method

F. Measuring Solute Potential

Solute potential can be measured by:

1. Freezing Point Depression - dissolved solutes lower the freezing point of a liquid (think salt and MN roads in the winter). A 1 molal solution with an osmotic potential of -2.27 MPa lowers the freezing point (fp) by -1.86 degrees. Thus: y_s = -2.27/1.86 and rearranging, y_s (MPa) = -1.22 (fp);
2. Incipient Plasmolysis - a tissue is incubated in a series of solutions of known water potential. The point at which membrane just pulls away from tissue is "incipient plasmolysis" and considered to the equivalent to the osmotic potential of the tissue.
3. Vapor pressure osmometer - dissolved solutes increase the boiling point or decrease the vapor pressure of a liquid. A thermocouple hooked to a recorder is placed in an airtight chamber with the tissue sample or standard. The thermocouple is also linked to a reference junction. Thermocouples are made of two different metals (constantan and chromel) and a current will flow if there is a difference in temperature between the reference junction and thermocouple. A water droplet or KCl (aq) solutions of known osmolarity are placed on the thermocouple. Depending on the osmotic potential of the solution, water will either evaporate from or condense on the droplet. This is turn causes a current change in the thermocouple which can be detected by a meter. The cooling rate is plotted vs. y_s or [KCl] to yield a standard curve from which the y_s of the tissue is determined.

The Movement of water across a membrane is a combination of diffusion and bulk flow

Individual water molecules diffuse across the membrane. In addition, there are integral proteins in the membrane that form a channel or pore through which water moves. These pores are important and water molecules essentially move through these pores by bulk flow. The proteins are called aquaporins and are essentially water transport channels. Water is moving passively (following a gradient of free energy).

Water Transport

I. Soil-plant-air continuum

The movement of water follows the pathway:

soil → uptake → root → stem → leaf → transpiration → air

The driving force for water movement is the water potential gradient that exists from soil to air. Or in other words:

Ψsoil > Ψroot > Ψstem > Ψ leaf > Ψair

Some typical values water potential values (in MPa) for a tree are: trunk -0.7; twig -2.3, leaf -2.5. In class, we may also look at some data for ivy.

II. Soil → Plant

A. What is soil?

Soil is a mixture of organic (dung, decayed organic materials, decomposers) and inorganic (weathered rock) materials, gases (oxygen, carbon dioxide, ethylene), and liquid.

B. Soil type

determined by: (1) composition; (2) texture or particle size (sand > silt > clay. A loam is a soil with 10-25% clay and equal parts of sand and silt); and (3) structure (*i.e.*, compaction)

C. Water and soil

1. Saturated - soil before drained. Gravitational water - water that drains and is not tightly bound; Ψ = 0 MPa
2. Field capacity - soil that holds all the water it can against gravity. Capillary water -water held by capillary action, water at field capacity; Ψ = -0.015 MPa

3. Permanent wilting percentage - soil moisture content at which plants can't get enough water. For most, Ψ = -1.5 MPa
4. Graphic relationship of soil water potential vs. water content (%). Take-home lessons
   a. between PWP and FC is the water available for a plant to use;
   b. clay holds more water than sand at any $\rightarrow\Psi$; and
   c. clay holds water more tightly (*i.e.*, @10% water Ysand > Ψclay). This is essentially a s/v problem, since smaller particles in clay they have a larger total surface and hence, has more charged surfaces that will bind water tightly.
5. Soil water potential is a function of osmotic potential (which is usually near zero except in saline soils) and mostly pressure (used to call it matrix potential; this refers to the tension generated because of the attraction of colloidal particles; *i.e.*, adhesion). The pressure in the soil can be calculated from the equation: $\Psi p = -2T/r$ *where* T = surface tension (7.28×10^{-8} MPa) and r = radius of curvature of the meniscus). Water movement through soil - mostly due to bulk flow as a result of pressure gradients, with some diffusion.
6. Spuds McSaupe plays with sponges

III. Plant $\rightarrow$ Air (= Transpiration)

Or more simply stated, the movement of water from plant to air occurs via transpiration. Air has a very high capacity for holding water. For example at 20 C, the water potential of water in air at 100% RH = 0 MPa; 98% RH -2.7 MPa; 50% RH = -93.5 MPa. Conclusion - there is a very steep water potential gradient from soil to air. Essentially, the plant just inserts itself between the two and takes advantage of passive transport.

IV. Soil $\rightarrow$ Root

A. Root anatomy

   We will go over structure of the root including epidermis, cortex, endodermis, Casparian strip, stele, phloem, xylem, pericycle.

B. Root formation

   Roots develop from the pericycle; film loop

C. Apoplast vs. symplast

   Recall that the apoplast refers to the "non-living" regions of the plant and the symplast is the "living" areas

D. Region of water absorption

Most of the water is absorbed near the tip of the root. The further from the tip, the less water that is taken up by the root. This roughly correlates to the region of the root that is suberized.

E. Route of water movement

There are three routes water can follow: (a) Apoplastic – water follows an apoplastic route from soil through cortex. However, it must enter the stele symplast because of the **Casparian strip**. Once inside, it leaks back out and enters the apoplast (xylem) where it is transported to the apex of the plant. This appears to be the major route of transport; (b) Symplastic: Transmembrane – in other words, the water moves from cell-to-cell crossing membranes as it goes; and (c) Symplastic: Plasmodesmata – the water moves from cell-to-cell via the plasmodesmata.

F. Root as an osmometer

Analogy - allows for the development of root pressures in the stem. These can be measured and are about 0.2 - 0.3 MPa.

G. Guttation and hydathodes

V. Root → Leaves

A. What is the transport tissue for water? Xylem.

Evidence comes from various tracer studies where xylem is loaded with dyes. I'll bet you've done the classic "celery stalk in food coloring" experiment.

We'll see a film loop and maybe play with some celery

B. In which cells does water move? Vessels & Tracheids

There are four major types of cells in the xylem: (a) tracheids - long, tapered ends, thick secondary wall; (b) vessel elements, - shorter, ends attached; (c) fibers - long and skinny with thick secondary wall, mostly for support; and (d) parenchyma - alive, thin, store starch and other materials, lateral transport. The primary water transport cells are tracheids and vessels. Note that gymnosperms only have tracheids whereas angiosperms have both and primarily rely on vessels for water transport. Both tracheids and vessels have pits, thin circular regions, in the walls.

C. How much pressure is required to move water to the top of a tall tree, that is say, 100 meters tall?

Let's calculate. We can measure the velocity of flow in the xylem to be 4 - 13 mm s^{-1} in vessels with a diameter of 100 - 200 mm. For our calculations, let's use a flow rate of 4 mm s^{-1} (= 4 x 10^{-3} m s^{-1}) and a vessel radius of 40 ìm (= 0.00004 m).

According to Poiseuille's Law - flow rate is directly proportional to the pressure gradient and the cross sectional area of the pipe but inversely proportional to the viscosity of the fluid. Thus, this is mathematically expressed as the Poiseuille equation:

$Jv = ((\pi)(r^4)(\Delta P))/8(\eta)$ where η = viscosity of water (assume it is the same as in a cell, 10^{-3} Pa s)

Now, divide the equation by the cross-sectional area of a vessel (πr^2). Thus, the equation simplifies to:

$Jv = ((r^2)(\Delta P))/8(\eta)$

Substituting the values for flow and vessel diameter:

4×10^{-3} m/sec = $(0.00004 \text{ m})^2(P)/8(10^{-3} \text{ Pa s})$

P = 20,000 Pa m^{-1}

P = 0.02 MPa m^{-1}

If the tree is 100 meters, then: 0.02 MPa m^{-1} x 100 m = 2 MPa

However, we must also take into account the effects of gravity (0.01 MPa m^{-1}). Thus, for a 100 m tree: 0.01 MPa m^{-1} x 100 m = 1 MPa for gravity

Finally, the total pressure required to move water to the top of a 100 meter tree equals:

2.0 + 1.0 = 3 MPa

D. How is Water Moved to the top of trees?

1. Is water moved to the tops of trees by a "push from the bottom" pump?

NO

Several lines of evidence show that this type of pump doesn't exist: (a) dissections showed there is no anatomical area in the stem or root that could serve as a pump; (b) when German physiologists cut off a tree in a vat of picric acid it continued to transport water. This suggested that a stem pump was not involved since the picric acid should have killed living cells stopping the pump; (c) a root pump isn't involved or else when a plant is decapitated, the stump should continue to gush water; and (d) recall that root pressures only generate

0.2-0.3 MPA but that a pressure of at least 3 MPa is required to move water to the tops of tall trees. To summarize, root pressure doesn't have nearly enough power.

2. Is water moved to the tops of trees by "capillary action" -

NO

Capillary action is the movement of water up a thin tube due to surface tension and the cohesive and adhesive properties of water. Essentially the meniscus "pulls" the water up the tube. Without worrying about the derivation of the equation, the height to which a column of water can move is inversely related to the radius of the pipe and is mathematically expressed as: h = 14.87/r (where r = radius in ìm; and h = height in meters). Let's look at some actual data

Table 1: Capillary Heights of Water Movement

Tube Radius (ìm)	Column Height (m)
10	1.4877
40 (tracheid)	0.37
100	0.148
0.005 (size of pores in wall)	2975 (*ca.* 3 kilometers)

Vessels are too wide to support movement very high and obviously capillarity cannot be responsible for water movement. Further, even it could, it would only move to the top of the plant once; capillary action can't continually pull the water up.

3. Cohesion-Tension Theory -

YES!

This idea was first proposed by HH Dixon (*Transpiration and the Ascent of Sap in Plants*, 1914). According to this hypothesis, water is drawn up and out of the plant by the force of transpiration. Because of the cohesive/adhesive properties of water, as one water molecule evaporates at the opening it pulls the other molecules and sends this pull all the way down the column. If this is true them, water transport in plants must meet the following criteria:

- *The system must have little resistance.* The vessels and tracheids are hollow at maturity. Imagine how difficult it would be to move water through a clogged pipe. Let's calculate how much pressure would be necessary if the water transport cells were "alive." We'll use Fick's Law:

Jv = Lp $\Delta\Psi$ (Lp = hydraulic conductance which is the inverse of resistance)

If we assume that water movement occurs at the rate we used earlier (Jv = 4×10^{-3} m s^{-1}) and we use a typical value of 4×10^{-7} m s^{-1} MPa^{-1} for Lp, then:

4×10^{-3} m s^{-1} = (4×10^{-7} m s^{-1} MPa^{-1}) (ÄØ)

ÄØ = 4×10^{-3} m s^{-1}/4×10^{-7} m s^{-1} MPa^{-1}

ÄΨ = 10^4 MPa (this is the pressure required to move the water across just one membrane! Compare to the value we calculated above)

if, the cell length is 100 ìm (10^{-4} m), then we can calculate the force required per meter:

$$10^4 \text{ MPa}/10^{-4} \text{ m} = 10^8 \text{ MPa m}^{-1}$$

- *The columns of water must be continuous from the leaves to the soil.*

  If not, it would be analogous to having a chain with a single broken link – it would be impossible to pull anything attached to the other end. The tracheids and vessels form a continuous water column. If there are gaps - or air bubbles - water must be routed around these bubbles. Cavitation, or vacuum boiling, is the fancy term for air coming out of solution when the water columns break.

  By the way, this is one reason why you don't want to go outside and beat on the trunk of a tree on a hot sunny day...it could cause many of the columns to break so that a plant may have a difficult time transporting water.

  If cavitation does occurs, the plant responds by: (a) transporting water around the blocked cell; or (b) redissolving the air bubble, which usually occurs at night; and/or (3) forming new xylem cells; in other words, xylem is disposable. Only the most recent cells in the latest seasons growth are actually functional. The remainder of wood in a tree is non-functional because it has cavitated and/or filled with other waste materials. In addition, it is thought that one function of bordered pits is to stop the movement of air bubbles from a cavitated cell to another thereby isolating the impact of cavitation.

- *There must be sufficient pulling force.*

  Even though *ca.* 3 MPa are required to move water to the top of a tall tree, the water potential gradient from soil to air is considerably steeper (on the order of -100 MPa.)

- *The xylem should be under a tension.*

  Several lines of evidence support this prediction: (a) Cut a stem and the water will snap up into the top and accumulate at the cut surface on the bottom; (b) Dendrometer studies - this device is essentially a band wrapped around a tree that is hooked to a pressure transducer. As the tree transpires the diameter of the tree is measured. These experiments show that the diameter of the stem is smallest during the day when transpiration is occurring and largest at night, as we would. Imagine putting your finger on the end of a straw and then sucking on the other end. The straw will get thinner (collapse) as you apply tension to the air in the straw - just like a plant stem; (c) Puncturing the xylem of an actively transpiring tree with an ice pick may result in a hissing sound as air is sucked into the stem; and (d) Dye solutions are rapidly sucked into a tree trunk when punctured with a knife and then transported in both upward and downward directions. Since the pressure in the stem is lower than atmospheric the dye solution is quickly sucked in. We will see a demonstration of this in a video that a previous class made (BIOL327- Spring 2001) made.

- *The tensile strength of water must be able to withstand the pull.*

  In other words, the columns of water must not snap as they are being pulled. As the water is pulled up the tree the water column puts up a resistance, much like stretching a rubber band. Just like a rubber band will snap if pulled too hard, so too will a water column break or cavitate. This occurs because the reduced pressures in the water column cause gases to come out of solution and form a vapor lock in that cell. Cavitation can be heard by placing a sensitive microphone on the plant. Tiny popping noises can be heard, a little like a bowl of rice krispies.

  The fact that water has a very high tensile strength, more than sufficient to withstand the pulling forces necessary to move water to the top of tall trees, was demonstrated by an elegant experiment in which water was centrifuged in Z-shaped tubes.

  As an aside, the tensile strength of water may be one of the factors that limits the height of trees - the tensions in the stems of taller trees would be too great and the water columns would snap. It's perhaps not a surprise that the tallest trees, California redwoods, grow along the fog enshrouded coast. This helps to minimize the rate of water loss and ultimately reduce the tensions in the xylem (Zimmer, 2000).

- *Tracheids and vessels must be able to withstand tensions without imploding.* Hence the reason that they have thick cell walls with circular thickenings. It's no surprise that wood is hard.

4. Pressure Compensation Theory - Controversial.

   Recent work by Martin Canny and others have challenged the validity of the Cohesion-Tension Theory.

VII. Flow/size paradox (compromise)

Don't you just love a paradox?....recall that flow rate is directly related to the radius of the pipe (Poiseuille's law). Thus, flow rates in vessels are greater than those in tracheids. But, why aren't vessels (and tracheids for that matter) larger, especially since it means that they could transport more water? The answer - cavitation. As the pipes get larger, the chance of cavitation increases.

CHAPTER - 25

Transpiration / Gas Exchange

I. Definitions

- Transpiration - evaporation of water from a plant surface
- Evapotranspiration - evaporation of water from a plant surface and the soil (and abiotic surroundings.
- Take-home-lesson: Plants loose a lot of water by transpiration. See text for some specific examples.

II. Photosynthesis/Transpiration Paradox (or perhaps more accurately, a "Compromise" or "Dilemma")

Recall the equation for photosynthesis where:

$CO_2 + H_2O$ 7 $(CH_2O)_n + O_2$

This equation suggests that:

1. Gases, such as carbon dioxide and oxygen, are important in the overall energy metabolism of plants;
2. Plants must exchange gases with the environment; and
3. In order to obtain carbon dioxide plants will necessarily loose water (transpire). In other words, transpiration is a necessary evil of photosynthesis.

III. Theoretical Considerations

A. A large surface area is required for efficient gas exchange.

Thus, animals have lungs or gills; plants have leaves and within the leaf - spongy mesophyll. It's time to review your leaf anatomy.

B. A large surface area for exchanging gases offers a large surface area for desiccation.

Animals solve this problem by placing the absorptive surface inside a humid cavity (lung) opened with a small exit pore(s). Plants put the absorptive surface (spongy mesophyll) inside the leaf and cover it with a water impermeable layer (cuticle) peppered with a series of pores (stomata). The cuticle is comprised of waxes, which are an assortment of long chain hydrocarbons and, in particular, cutin (C16-C18 hydroxylated fatty acids). Note that even though these strategies minimize desiccation from the absorptive surfaces of plants and animals, they don't completely eliminate it.

C. Placing the absorptive surfaces inside the organism to reduce desiccation presents a problem - getting the gases to the absorptive surface.

Animals use an active pumping mechanism (*e.g.*, lungs/diaphragm) to move gases inside the organism by bulk flow. The gases are circulated by another pumping system (heart). The distances needed to move the gases are too great to be accounted for by simple diffusion. Plants do not have a pumping mechanism for moving gases; they rely on diffusion (and bulk flow). In either case, plants do not actively move gases. This is one reason why leaves are so thin (recall our previous discussion) - diffusion is not efficient over long distances (*i.e.*, diffusion is inversely related to the square of distance). For example, it would take about 2.5 seconds for glucose to diffuse the distance across a cell membrane (0.5 μm) but approximately 32 years to go one meter!

D. Paradox and Compromise.

In order to obtain carbon dioxide for photosynthesis plants needed to evolve a large, thin absorptive surface (leaves with spongy layer) and then protect it from desiccation. Thus we can consider this the photosynthesis/transpiration "paradox". Actually, it might better be considered a "compromise" since that is what a plant needs to do - strike a compromise or balance between the amount of carbon dioxide absorbed and the amount of water lost by transpiration.

IV. Further Complications

Not only is water loss a "necessary evil" of photosynthesis, but to make matters worse, the tendency to loose water is greater than the tendency of carbon dioxide to diffuse into the plant. As evidence, let's calculate the transpiration ratio, which is a measure of the the amount of water loss relative to the amount of carbon fixation.

transpiration ratio = mol water transpired / mol CO_2 fixed

If carbon dioxide uptake (or fixation) and water loss are equal, this ratio should be one. In reality, experiments show that this ratio is closer to 200! Thus, for every 200 kg of water transpired, 1 kg of dry matter is fixed by a plant. Let's see why:

A. Diffusion and gradients

Recall that during diffusion molecules move from an area of higher chemical potential (or concentration or chemical energy) to an area of lower chemical potential (or concentration or free energy) and that the driving force for diffusion is the gradient from one area to another. We can express this relationship mathematically using Fick's Law: $Jv = (c_1 - c_2)/r$ *where* Jv = flux density, (mol $m^{-2}s^{-1}$); $c_1 - c_2$ = concentration gradient, and r = resistance (a function of distance, medium viscosity, membrane permeability, etc.). We can simplify this equation to:

diffusion = gradient/resistance (distance)

Now let's compare the rates of diffusion for both water and carbon dioxide. Since resistance, or the distance that either carbon dioxide or water must diffuse into/out of the leaf is the same, then diffusion is directly proportional to the gradient.

1. Carbon dioxide - has a very shallow, or small, gradient from inside to outside of the leaf. Ambient carbon dioxide concentrations are approximately 0.03% (= 0.34 mmol mol^{-1}) and the internal concentration cannot be less than zero. Thus, the gradient is no larger than 0.34 mmol mol^{-1} (0.34 - 0), a relatively small difference.

2. Water - has a very steep gradient from inside to outside. At a relative humidity of 50% and 25 C the water potential of water in air is *ca.* -100 MPa (=32 mmol mol^{-1}). The air in the substomatal cavity of a leaf is typically fully saturated, with a RH near 100%, and has a water potential near zero. Thus, the water potential gradient is 100 MPa (or 32 mmol mol^{-1}), which is considerably steeper than that for carbon dioxide.

Conclusion: based on gradient alone, the water has a much greater tendency to diffuse out of the leaf than carbon dioxide to diffuse into the leaf.

B. Diffusion and molecular weight

Recall that diffusion is inversely related to molecular weight. Simply put, the heavier the molecule the more slowly it will diffuse. No big surprise. Or, to express this mathematically:

$$\text{rate} = 1 / \text{sq rt MW}$$

The relationship between the rate of diffusion of two molecules can be summarized by the following relationship:

$$\text{Rate A} / \text{Rate B} = \text{sq rt B} / \text{sq rt A}$$

Thus to calculate the ratio of water loss to carbon dioxide uptake:

$$H_20 \text{ loss}/CO_2 \text{ uptake} = \text{sq rt } 44 / \text{sq rt } 18 = 1.56$$

Conclusion: Based on molecular weight alone, the tendency for water to diffuse out of the leaf is 1.56 times greater than the tendency for carbon dioxide to diffuse into the leaf.

V. The Photosynthesis/Transpiration Compromise Revisited

Although it seems as though water loss is a serious, intractable problem for a plant, it is NOT. The reason - plants continually compromise between the amount of carbon dioxide absorbed and the amount of water loss. This compromise is mediated by the stomata, whose function is to regulate gas exchange.

VI. Stomatal Structure

A. *Anatomy of a stoma (stomata, plural)* - guard cells, subsidiary cells, substomatal cavity, cuticle, ledge (or lip), stomatal apparatus. The subsidiary cells are epidermal cells that may be specialized and different from the other epidermal cells. The function of the ledge is to prevent liquid water from seeping into the pore. Interestingly, cutin covers most of the cells in the substomatal cavity; only regions near the actual opening are free of cuticle and most water is lost from this area.

B. *Types of guard cells:* (1) elliptical or kidney-shaped. These are characteristic of dicots and other non-grasses, and (2) dumb-bell or dog-bone shaped - characteristic of grasses (called graminaceous type)

C. *Common features* - (1) thickened inner walls; and (2) radial micellation - the cellulose microfibrils radiate out around the circumference of the pore; (3) chloroplasts - these are the only epidermal cells with chloroplasts; and (4) connected end-to-end.

VII. The Beauty of Stomata

The evolution of a water-impermeable covering of the absorptive surface that was peppered with oodles of pores was a great idea. The stomata are ideal structures for regulating gas exchange because:

A. *There are lots of them on any plant surface.* In fact, there can be as many as 1000 mm^{-2}. Obviously, the larger the number of pores, the greater the amount of total diffusion that can occur . A large number of pores is necessary so that plants are able to absorb enough carbon dioxide for photosynthesis.

B. *The pores comprise a large area of the surface of a plant.* I n fact, stomata occupy as much as 2-3% of the total leaf surface area. As expected, the greater the pore area the greater the rate of diffusion which is advantageous for maximizing carbon dioxide uptake.

C. *The pores are small.* On average, stomata are about 14 ìm in diameter. Although more total diffusion occurs through large pores [*total diffusion (ìg* hr^{-1}*) vs. pore dia. (ìm)*], small pores are more efficient than larger ones. This is due to the edge effect (related to surface-to-volume ratio). Smaller pores have a greater proportion of edge. As molecules reach the edge they "spill over" and in effect have a shorter diffusion distance to get away from the pore.

D. *The pores are optimally spaced.* A significant boundary layer of humid air forms around leaf surfaces. This humid area reduces the rate of further transpiration. It occurs because diffusion shells from adjacent pores fuse. If the pores were much farther separated, they wouldn't form a nice boundary layer. The thickness of the boundary layer is further affected by: (1) wind speed; (2) presence of hairs; and (3) sunken chambers.

E. *The pores are optimally located.* In grasses, the stomata are distributed approximately equally on both sides of the leave whereas in herbaceous dicots there are generally more on the underside (abaxial) than the upper (adaxial) side. In woody dicots there are usually few stomata in the upper surface, whereas aquatic plants with floating leaves have most stomata in the upper surface. These modifications are important to minimize/control water loss. Conifers and xeric plants often put the stoma in sunken chambers.

F. *The degree of opening/closing of the pore is closely regulated by the plant in response to the environment.* In effect, any factor that can impact the rate of photosynthesis or overall water status of the plant will influence the action of the guard cells (see below). I like our textbook description that stomata are "multisensory hydraulic valves."

G. Test these ideas by studying the data supplied

VIII. Mechanics of Guard Cell Action

Guard cells open because of the osmotic entry of water into the GC. In turn, this increases the turgidity (water pressure) in the GC and causes them to elongate. The radial orientation of cellulose microfibrils prevents increase in girth. Since GC are attached at the ends and because the inner wall is thicker, the guard cells belly out with the outer wall moving more pulling open the guard cell. Guard cell closure essential involves reversing this process.

In class we'll see a great, old, film loop about this process. We can summarize the mechanics of GC action as follows:

stoma closed *(GC flaccid)* 7 water uptake (osmosis) 7 increase pressure 7 stoma open *(GC turgid)*

IX. Physiology of Guard Cell Action. Part I.

Since water is the driving force for GC action, this means that there must be a gradient in water potential between the GC and the surrounding cells (subsidiary cells). Thus, to open a stoma, there must be a mechanism to generate a water potential gradient.

A. Hypotheses for how the water potential gradient is established include:

1. There is a rapid decrease in pressure in surrounding cells (i.e., subsidiary cells shrink which would result in the GC expanding and taking up water). Using a fine needle transducer, Mary Edwards and Meidner showed that there is some decrease in surrounding cell P, but this is not a major factor.
2. There is an increase in the stretchability of the GC cell wall. This would result in an expansion of the GC with a concomitant uptake of water. Little evidence exists for this idea.
3. There is a decrease in the osmotic potential in the GC - Much evidence supports this hypothesis. For example, Humble and Raschke (1971)

showed that the solute potential of turgid GC (open stoma) of broad bean is -3.5 MPA but when the guard cells are flaccid (stoma closed), the osmotic potential is -1.9 MPa. Thus, the solute potential of the GC decreases (becomes more negative) when open. Since the GC volume increases, this must mean that there is an accumulation or synthesis of solute.

Thus, we can modify our schematic diagram:

stoma closed (GC flaccid) ❼ add solute ❼ lower Ψs ❼′ decrease ψw

❼ water uptake (osmosis) ′ increase pressure ′ stoma open (GC turgid)

B. *What is the solute and where does it come from?*

1. Carbohydrates, such as sucrose

   Since guard cells are the only epidermal cells with chloroplasts, plant physiologists have long hypothesized that sucrose and related carbohydrates are osmotic regulators of guard cells. For example, the starch content of guard cell chloroplasts decreases as the stomata open. This idea, the "starch-sugar hypothesis", was the first postulated mechanism for guard cell activity. It lost popularity after the role of potassium ions was discovered, but most now agree that both sugar and potassium ions play a role in guard cell regulation. Sucrose seems to be especially important in closing guard cells.

   Where does the sucrose come from? (a) hydrolysis of starch in the GC chloroplasts. In other words, an indirect product of photosynthesis (evidence: starch grains disappear during opening); or (b) a direct product of carbon fixation (photosynthesis).

2. Malate

   Malate is an organic acid (C4). You may already be familiar with its role in the Kreb's cycle in the mitochondria. In plants, malate is also derived from the hydrolysis of starches. The enzyme phosophoenolpyruvate carboxylase (PEPase) binds carbon dioxide (actually bicarbonate ions) to (PEP, 3-carbons, an intermediate in glycolysis) to produce oxaloacetate (C4) which is then reduced to malate and stored in the vacuole.

3. Chloride ions

   Chloride ions are transported into the cell from the apoplast via a Cl^- / H^+ symport in which a proton gradient is used to "drag" the chloride into the cell.

4. Potassium ions

   This appears to be the primary osmotic agent, especially for opening the GC in the morning. The potassium comes comes from surrounding cells. Evidence: (a) if you strip epidermis from a leaf, which breaks many epidermal cells but not the more resistant GC, the GC will only open if K^+ is provided in the medium; (b) potassium concentrations increase in the guard cells upon opening.

Table 1: Potassium in the stomatal aperture of *Commelina communis*

	Open	**Closed**
K^+ (mol) in GC	0.45	0.10
K^+ (mol) in epidermal cells	0.07	0.45

Thus, we can modify our original scheme:

stoma closed (GC flaccid) → sucrose/potassium/malate/chloride ions → lower ψs → decrease ψw → water uptake (osmosis) → increase pressure → stoma open (GC turgid)

To close the stoma, the reverse process occurs. **However**, time course studies indicate that potassium uptake is associated with opening of the stomata in the morning, but sucrose loss is more closely associated with closure in the afternoon. Thus, the final modification to our scheme:

stoma closed (GC flaccid) → potassium and chloride ion uptake, malate synthesis → lower ψs → decrease ψw → water uptake (osmosis) from subsidiary cells → pressure increases → stoma open (GC turgid) → ||||→ sucrose (potassium, chloride, malate) decreases → ψs increases → ψw increases → water loss → pressure decreases → stoma closed (GC flaccid)

X. Environmental Control of GC Action

Whatever physiological mechanism we finally postulate for the GC, it must also be compatible with the action of various environmental factors that are known to regulate stomatal activity. Since guard cells respond to their environment, especially any factors that impact the photosynthesis/transpiration compromise. We expect any factor important in photosynthesis to exert regulatory control on GC. And, we expect water to have the final word on control since if a plant dries out too much it's as good as dead!

A. Light - exerts strong control. In general: light = open; dark = closed *(except CAM plants)*.

What kind of light is important?

1. Red & blue light

   The action spectrum for the process suggests that both red and blue light are important regulators of guard cell activity and that their action is, at least partially, mediated by photosynthesis (recall red and blue light are used in photosynthesis). Further evidence that photosynthesis is important - DCMU (an inhibitor of PS) prevents stomatal action.

   So, what is photosynthesis doing? (a) provides sugars (sucrose and glucose) for osmotic regulation; (b) provides ATP (via photophosphorylation) to power ion pumps (see below); (c) reduces internal CO_2 levels which stimulates opening (see below); and (d) reduces the pH in the lumen of the thylakoid that stimulates the synthesis of the blue light receptor pigment (see below).

2. Blue light

   There is an additional effect of blue light on stomatal activity that is irrespective of its role in photosynthesis. The evidence: (a) blue light is 10x more effective than red light; (b) in saturating levels of red light, treatment with blue light will cause additional GC opening; (c) the action spectrum for blue light is similar to other "blue light responses", a "three-finger pattern"; (d) photosynthesis inhibitors block the red light effect but not the blue light effect.

What is blue light doing?

- Blue light activates a H^+-ATPase in the membrane (recall the proton pump for cell elongation?). Evidence: (a) potassium accumulates in isolated GC protoplasts treated with blue light and causes them to swell; (b) blue light causes the acidification of the medium of GC protoplasts under saturating red light conditions; (c) fusicoccin, which stimulates proton pumping, also stimulates stomatal action; (d) vanadate (VO_3^-, blocks the proton pump) and CCCP (carbonyl cyanide m-chlorophenylhydrazone, an ionophore that makes the membrane leaky to protons) both inhibit stomatal opening.
- Blue light stimulates starch breakdown and malate synthesis.
- Blue light stimulates cellular respiration (which among other things may be required to produce ATP for the proton pump.

What is the receptor for the blue light?

The action spectrum for the blue light response shows a "three finger pattern," which is characteristic of other blue light responses (i.e., phototropism). Absorption spectra of potential receptor pigments show a good match between zeaxanthin, a carotenoid pigment (C40) that occurs in the chloroplast thylakoid, and the action spectrum. Further – zeaxanthin levels are directly correlated with stomatal aperture.

How does blue light cause stomatal closure?

Photosynthetically active radiation (red and blue light) cause an acidification of the chloroplast lumen. This activates the synthesis of zeaxanthin, which in turn, zeaxanthin activates a calcium-ATP pump in the chloroplast membrane that decreases calcium concentrations in the cytosol. This, in turn activates the proton pump in the cell membrane.

B. Low oxygen levels → GC open

C. Carbon dioxide - intracellular level is an important regulatory control.

- lo CO_2 (*i.e.*, during the day, used by photosynthesis) = open
- hi CO_2 (*i.e.*, at night, produced during respiration) = closed

D. pH effect

- hi pH → open
- lo pH → closed

This effect seems mediated by:

1. Carbon dioxide Concentration.

   Recall the interaction of carbon dioxide and water:

   $CO_2 + H_2O \rightarrow CO_2\ (aq) \rightarrow H_2CO_3\ (aq) \rightarrow H^+ + HCO_3^- \rightarrow H^+ + CO_3^-$

   therefore:

   low CO_2 = hi pH = open

   hi CO_2 = lo pH = closed

2. H+/ATPase proton pump.

   The pump is required for stomatal opening (see above). Protons are transferred from the cytosol into the apoplast. As protons are removed from the cytosol, the pH increases.

E. Water - protects against excessive water loss.

This is the prevailing and overriding control mechanism. There are two mechanisms by which water loss regulates stomatal closure, one of which is active and the other passive.

1. Hydropassive Control - simply put, as the plant looses water, the turgidity of the leaf cells, including guard cells, decreases and this results in stomatal closure. The plant is not "intentionally" closing the stoma, it is simply a consequence of drying out.
2. Hydroactive Control - this mechanism is one in which the plant actually seems to monitor its water status. When the water potential drops below some critical level, it engages a cascade of events that close the stomata. Presumably the plant is measuring pressure (turgor) and then synthesizes or releases an anti-transpirant that is translocated (moved) to the GC to cause closure.

The anti-transpirant is abscisic acid (ABA), one of the major plant growth regulators. It is active in very low concentration (10^{-6} M) and appears very rapidly after water stress (within 7 minutes). After 4-8 hours, the [ABA] increases nearly 50x. ABA comes from two sources: (a) root – in response to water stress, the xylem sap pH increases which in turn stimulates the release of ABA into the xylem sap for transport to the leaves. This seems to be a root signal to the leaves that "water stress is coming"; and (b) leaves - water stress stimulates a synthesis of new ABA and redistribution of existing ABA.

Mechanism of Action

Treatment with ABA results in decrease of potassium, chloride and malate levels in the guard cells which in turn increase the water potential resulting in water efflux. Evidence suggests that there is an ABA receptor in the cell membrane. The receptor: (a) activates calcium channels in the membrane causing calcium uptake from the apoplast; (b) activates calcium channels in the tonoplast causing calcium release from the vacuole into the cytosol; (c) activates chloride (and malate) *efflux* channels; (d) inactivates potassium ion *"in"* channels; (e) inactivates the cell membrane proton pump; and (f) causes an increase in pH that activates potassium *efflux* channels. Thus, in short, ABA treatment causes an increase in cellular calcium levels which in turn results in decreases potassium and chloride levels and turns off the proton pump.

F. Temperature

Increased temperatures usually increase stomatal action, presumably to open them for evaporative cooling. If the temperature becomes too high the stomata close due to water stress and increased CO_2 that results from respiration.

G. Wind

Often causes closure because it: (a) brings CO_2 enriched air; and (b) increases the rate of transpiration that causes water stress which causes the stomata to close. In some cases, wind causing stomatal opening to increase transpiration for cooling.

XI. Physiology of Guard Cell Action - Part II

Now, let's pull everything we've learned together to hypothesize a mechanism of action. First, there are a couple of other observations that we also need to reconcile with our mechanism:

A. Starch disappears in open stomata.

Alkaline conditions favors the starch hydrolysis and acidic conditions favors starch synthesis. Starch hydrolysis is activated by blue light; and

B. PEPcase activity is high in GC.

Phosphoenolpyruvate carboxylase catalyzes the reaction: phosphoenolpyruvate (PEP) + HCO_3^- '! oxaloacetate (OAA).

XII. Why does Transpiration Occur?

A. Transport in plants. This is important to a small degree. Transpiration is certainly not a necessity.

B. Heat loss (latent heat of vaporization)

C. Carry nutrients in the soil to the plant

D. Perhaps plant cells need to maintain some optimal level of turgidity and this helps them do so.

CHAPTER - 26

Solute Transport: Phloem Structure and Function

I. Definition

Solute transport in plants, translocation, primarily occurs in the phloem, but it can occur in the xylem.

II. Solute Transport in the Xylem

- Some solutes are transported in the xylem
- Water and dissolved ions are the main substances in vessels/tracheids
- These materials are transported via transpiration stream
- Xylem sap may also contain organic materials, usually in relatively low concentration (with a notable exception being maple sap in the spring which is comprised of 2% or more sucrose).
- Substances move at different rates depending on matrix effects, metabolic needs, etc.

III. Solute Transport in the Phloem

A. Phloem is difficult to study in plants because: (1) the transport cells/ tissue in plants are small (microscopic) in comparison to the transport structures in animals; (2) there is a very rapid response of the phloem to wounding (contents under pressure); (3) transport in plants is intracellular (vs. extracellular in animals); and (4) the transport cells are alive.

B. Phloem is the primary transport tissue for photosynthates (photoassimilates, or simply stated - organic materials).

Radiotracer studies in which leaves are briefly exposed to ^{14}C-labeled carbon dioxide show that radioactive photosynthates are localized in the phloem.

C. Aphids Don't Suck

Kennedy & Mittler (1953) first noted that aphids could be used as a direct pipeline to the phloem. Phloem-feeding aphids stick their hollow, syringe-like stylet directly into phloem cells. Surprisingly, the phloem doesn't seal itself in response. Aphids don't suck; rather, the phloem contents are forced into the aphid (thus the phloem is under pressure) and the excess oozes out the anus (honeydew). Thus, aphid studies demonstrate that the phloem is under pressure. Further, the honeydew can be collected and we can identify its composition. Better yet, after anaesthetizing the aphid with CO_2 the body is severed from the stylet leaving a miniature spile tapped directly into the phloem.

D. Phloem Content

Analysis - early studies to determine the content of the phloem involved cutting into the plant and analyzing the contents of the sap that was recovered. The problem is that you couldn't be sure that your sample wasn't contaminated by xylem exudates or other materials. Aphid studies described above helped to solve this problem. Phloem is rich in:

1. Carbohydrates - make up 16-25% of sap. The major organic transport materials are sucrose, stachyose (sucrose-gal), raffinose (stachyose-gal). These are excellent choices for transport materials for two reasons: (a) they are non-reducing sugars (the hydroxyl group on the anomeric carbon, the number one carbon, is tied up) which means that they are less reactive and more chemically stable; and (b) the linkage between sucrose and fructose is a "high-energy" linkage similar to that of ATP. Thus, sucrose is a good transport form that provides a high energy, yet stable packet of energy;

2. Amines/amides (0.04-4%) such as asparagine, glutamine, aspartic acid, ureides like ureas, citrulline, allantoin and allantoic acid. These compounds serve to transport "nitrogen";

3. ATP, hormones, sugar alcohols like sorbitol (apple, pear, prune) and mannitol (mangrove, olive), and an assortment of other organic materials; and
4. Inorganic substances including magnesium and potassium.

E. Direction of phloem transport

(1) Classic girdling experiments (removing the bark of a woody plant) by Malphigi (1675) and Hales (1725) provided some of the earliest evidence. These experiments showed the accumulation of material above the girdle, and that carbohydrates were not translocated below the girdle. Thus, plants transport substances in the phloem downward toward the roots.

(2) Sophisticated girdling experiments, using tracers like ^{32}P, ^{13}C, and ^{14}C demonstrate that substances in the phloem are transported downward towards the roots ***OR*** upwards toward the shoot meristem.

(3) Aphids and tracers

Conclusions - phloem transports organic materials from sites of production (called a **source**) to a site of need (called a **sink**). Thus, the typical direction of transport is downward from the primary source (leaves) to the major sink (roots).

F. Rate of phloem transport

Aphid experiments once again provide an answer...translocation rates average about 30 cm $hour^{-1}$ or even faster.

G. The phloem is under pressure

Studies with aphids showed that the sap was "pushed" out of the plant suggesting the phloem is under pressure. More recent studies with sophisticated pressure probes have shown a pressure gradient from source to sink.

IV. Phloem Anatomy

A. Cell types.

1. Sieve tube members or sieve elements.

These cells are joined end to end to make a sieve tube. Theye are called sieve cells in gymnosperms. At maturity, these cells: (a) are alive, (b) have a functional plasma membrane and therefore are osmotically active/responsive; (c) no tonoplast or vacuole; (d) no nucleus, thus no DNA-directed protein synthesis, (e) few mitochondria or plastids; (f) the ER is primarily beneath plasma membrane and it is mostly smooth.

Sieve elements are joined by sieve plates. These have numerous pores lined with callose (β 1-3 glucan). Callose forms rings around the pore, like a grommet. The wall region in the middle of the grommet hollows out and the membranes from the two adjacent cells are connected. Callose can plug the pore if the cell is damaged. The amount of callose observed varies with season, age, metabolism. Callose synthase is in the cell membrane.

2. Companion cells (angiosperms; albuminous cells - gymnosperms)

   These cells have a dense cytoplasm, mitochondria, nucleus, golgi, ER, chloroplasts - the standard goodies. Although their function is not well understood, they can be considered "nurse cells" to the sieve tube members. These cells are derived from the same cambial initial cell as the sieve tube members. There are three types of companion cells:

   (a) "ordinary" - with chloroplasts, few plasmodesmata connections to other cells except sieve elements, smooth inner walls, normal chloroplasts;

   (b) transfer – more plasmodesmata, ingrowths in the wall to increase the S/V ratio; and

   (c) intermediary – many plasmodesmata, vacuoles, undeveloped chloroplasts. The transfer and "ordinary" companion cells likely function to remove solutes from the apoplast

3. Parenchyma cells

   These are vacuolated, storage cells. They help in lateral conduction and may help in transferring material to/from sieve cells. Transfer cells are specialized parenchyma cells.

4. Fibers - primarily for support.

5. Side note: Mesophyll cells in a leaf are close (perhaps 1-3 cells away) to a minor vein.

V. P protein

- MW 14,000-158,000
- Originally thought to be a carbohydrate and was called slime because it gelled when exposed to the air

- Various forms, bundles of fibers or amorphous areas or even crystalline
- Appear early in development of sieve elements
- Only in angiosperms
- at least two proteins, PP1 and PP2
- Once the sieve pores form, the P-protein disperses through the pore.
- The protein is fibrous
- P protein plugs the pore when the cell is damaged.
- synthesized in companion cells

VI. Mechanism for Phloem Transport

A. Requirements

The model must account for: (1) speed of transport. The process is much faster than simple diffusion. For example, a conservative estimate of the mass transfer rate in phloem is 15 g cm^{-2} hr^{-1}. If the rate was based solely on diffusion is would be predicted to be 200 μg cm^{-2} hr^{-1}; (2) bidirectional flow - recall that substances can be transported down or up in the phloem; and (3) pressures in the phloem

B. Pressure flow (or Bulk Flow) hypothesis of Munch. This is the best model that fits the data.

The Model: Phloem transport is analogous to the operation of a double osmometer *(see diagram)*. If solute is added to bulb A → osmotic potential decreases → osmotic uptake of water → pressure increases → bulk flow of water and solute to bulb B → pressures increases in bulb → water potential in B greater than in beaker → osmotic flow of water into the beaker → water returns to side A via the connection. This system could be maintained indefinitely if there is a mechanism to remove solute (sucrose) at the end (sink) and a mechanism to add solute (source).

Sinks include young leaves, roots, developing fruits. Sources include mature leaves, cotyledons, endosperm, and bulbs and storage roots in spring. Sinks and sources can change depending upon the nutritional need of the plant. Thus, roots can be a source in the spring but are sinks for the majority of the growing season.

C. Plants as osmometers. If this model is valid then.....

1. Sieve tubes should be continuous pipes...they are.

2. Sieve tubes should provide minimal resistance to flow.

   In other words, the sieve tubes shouldn't be clogged by P-protein. This is true in specimens that are rapidly prepared. However, this was a major concern in early experiments because the phloem always appeared clogged up in TEM pictures. Further, this explains why sieve tube members have few "typical" cellular structures - they would "get in the way."

3. The phloem should be under pressure. As the aphid experiments suggest....it is.

   In fact, mini-pressure gauges can be attached to a severed aphid stylet and the pressure can be measured. It varies from 0.1-2.5 MPa. Further, there should be a pressure gradient from source to sink (driving force for movement). There is...see overhead.

4. Sieve elements must have a membrane (for development of pressure gradients) - they do.

5. There should be an osmotic potential gradient from source to sink (there is...see overhead).

   The source region of the phloem has a considerably lower osmotic potential than the sink regions.

6. There must be a mechanism to load solutes from the source into sieve cells.

   This process must be active since the solutes (usually sucrose) are being loaded against a concentration gradient. Evidence - respiratory inhibitors block the process. The loading mechanism should be:

   - Selective - it should only load the materials that are transported. This is supported by radiotracer studies; abraded leaves have been shown to only load materials that are normally transported;
   - Allow for apoplastic (from protoplast to wall to protoplast) or symplastic (from protoplasts to protoplast via plasmodesmata) transport. In some species, sucrose transport is symplastic - from mesophyll protoplast to cc-se protoplast via plasmodesmata. In others, sucrose loading into the cc-se complex involves an apoplastic step (mesophyll protoplasts to apoplast to cc-se protoplast.
   - Provide a mechanism to transport sucrose across the membrane - the sucrose/proton cotransport system. According to this model, protons are pumped out of the sieve cells into the apoplast by a

membrane-bound H^+-ATPase → the proton concentration increases in the apoplast → pH decreases → K^+ is brought into the sieve cell to balance the charge → the proton gradient provides the driving force for transporting sucrose against a gradient → the sucrose and protons bind to a carrier protein in the membrane and are released in the sieve tube member. Evidence: the pH is high in sieve tubes; if the pH of the apoplast is increased there will be no sucrose uptake; there is a hi potassium conc. in sieve tube members. A membrane carrier is likely involved since PCMBS (p-chloromercuribenzene sulfonic acid), an inhibitor of membrane proteins, interferes with sucrose uptake.

7. There must be a mechanism to unload solute at the sink. Sucrose is unloaded into the apoplast in some tissues (*i.e.*, ovules) and into the symplast of others (growing/respiring tissues like young leaves, meristems).
8. Apoplastic transport and unloading can occur via two methods: (a) sucrose is hydrolyzed by acid invertase to glucose and fructose upon reaching the sink. This maintains the gradient for transport. The glucose and fructose are taken up by the sink cells and stored or further metabolized as in maize; or (b) sucrose is unloaded into the sink by a carrier co-transport system like in sucrose loading.
9. The empty ovule technique has been useful in these studies.
10. Some metabolism is required (for loading/unloading) and to maintain sucrose against a concentration gradient. This explains the response to respiratory inhibitors. Phloem transport is also inhibited by anoxia and cold temperatures - both thought to exert their effect through energy metabolism.

VII. Problems with the Model

Bidirectionality - how can phloem translocate materials in two different directions at once? It can't, at least not within the same sieve tube. However, presumably sieve tubes within a single vascular bundle could be transporting in opposite directions assuming each is acting appropriately.

CHAPTER - 27

Mineral Nutrition in Plants

PLANT NUTRITION

Plants use inorganic minerals for nutrition, whether grown in the field or in a container. Complex interactions involving weathering of rock minerals, decaying organic matter, animals, and microbes take place to form inorganic minerals in soil. Roots absorb mineral nutrients as ions in soil water. Many factors influence nutrient uptake for plants. Ions can be readily available to roots or could be "tied up" by other elements or the soil itself. Soil too high in pH (alkaline) or too low (acid) makes minerals unavailable to plants.

ESSENTIAL MINERAL ELEMENTS

Total 112 elements have been discovered until now. So you might be wondering whether plants require all 112 elements for mineral nutrition of them. Most of the mineral elements present in soil are absorbed by roots of the plant. But all are not essential. Only 17 elements are considered as essential for the plants.

Sources of Essential Elements

Elements	Sources of the Elements
Carbon	Taken as CO_2 from the atmosphere (air)
Oxygen	Absorbed in the molecular form from air or from water. It is also generated within a plant during photosynthesis.
Hydrogen	Released from water during photosynthesis in the plant
Nitrogen	Absorbed by the plants as nitrate ion (NO_3^-) or as ammonium ion (NH_4^+) from the soil. Some like bacteria and cynobactreria can fix nitrogen from air directly.

Contd...

Potassium, calcium, iron, phosphorus, sulphur, magnesium	absorbed from the soil (are actually derived from the weathering of rocks. So they are called mineral elements.) They are absorbed in the ionic forms e.g. K^+, Ca^{2+}, Fe^{3+}, $H_2PO_4^-/HPO_4^{2-}$ etc.

Criteria for Essentiality of Elements

The nutrients or elements which are essential for the healthy growth of the plant are called essential nutrients or essential elements. The roots absorb about 60 elements from the soil. To determine which one is an essential element, the following criteria are used :

(i) An essential element is absolutely necessary for normal growth and reproduction of the plant.

(ii) The requirement of the element is very specific and it cannot be replaced by another element.

(iii) The element must be directly involved in plant metabolism.

Example : Magnesium is said to be an essential element because it is necessary for the formation of chlorophyll molecule. Its deficiency causes yellowing of leaves.

These criteria are important guidelines for plant nutrition but exclude beneficial mineral elements.

BENEFICIAL ELEMENTS

Beneficial elements are those that can compensate for toxic effects of other elements or may replace mineral nutrients in some other less specific function such as the maintenance of osmotic pressure. The omission of beneficial nutrients in commercial production could mean that plants are not being grown to their optimum genetic potential but are merely produced at a subsistence level. This discussion of plant nutrition includes both the essential and beneficial mineral elements.

CATION EXCHANGE CAPACITY (CEC)

The Cation Exchange Capacity refers to the ability of the growing medium to hold exchangeable mineral elements within its structure. These cations include ammonium nitrogen, potassium, calcium, magnesium, iron, manganese, zinc and copper. Peat moss and mixes containing bark, sawdust and other organic materials all have some level of cation exchange capacity.

ACIDITY AND ALKALINITY IN SOIL

The term pH refers to the alkalinity or acidity of a growing media water solution. This solution consists of mineral elements dissolved in ionic form in water. The reaction of this solution whether it is acid, neutral or alkaline will have a marked effect on the availability of mineral elements to plant roots. When there is a greater amount of hydrogen H+ ions the solution will be acid (<7.0). If there is more hydroxyl OH- ions the solution will be alkaline (>7.0). A balance of hydrogen to hydroxyl ions yields a pH neutral soil (=7.0). The range for most crops is 5.5 to 6.2 or slightly acidic. This creates the greatest average level for availability for all essential plant nutrients. Extreme fluctuations of higher or lower pH can cause deficiency or toxicity of nutrients.

THE ELEMENTS OF COMPLETE PLANT NUTRITION

The following is a brief guideline of the role of essential and beneficial mineral nutrients that are crucial for growth. Eliminate any one of these elements, and plants will display abnormalities of growth, deficiency symptoms, or may not reproduce normally.

Macronutrients

Macronutrients can be broken into two more groups:

primary and **secondary nutrients**.

The **primary nutrients** are nitrogen **(N)**, phosphorus **(P)**, and potassium **(K)**. These major nutrients usually are lacking from the soil first because plants use large amounts for their growth and survival.

The s**econdary nutrients** are calcium **(Ca)**, magnesium **(Mg)**, and sulfur **(S)**. There are usually enough of these nutrients in the soil so fertilization is not always needed. Also, large amounts of Calcium and Magnesium are added when lime is applied to acidic soils. Sulfur is usually found in sufficient amounts from the slow decomposition of soil organic matter, an important reason for not throwing out grass clippings and leaves.

Nitrogen is a major component of proteins, hormones, chlorophyll, vitamins and enzymes essential for plant life. Nitrogen metabolism is a major factor in stem and leaf growth (vegetative growth). Too much can delay flowering and fruiting. Deficiencies can reduce yields, cause yellowing of the leaves and stunt growth.

Phosphorus is necessary for seed germination, photosynthesis, protein formation and almost all aspects of growth and metabolism in plants. It is essential for flower and fruit formation. Low pH (<4) results in phosphate being chemically locked up in organic soils. Deficiency symptoms are purple stems and leaves; maturity and growth are retarded. Yields of fruit and flowers are poor. Premature drop of fruits and flowers may often occur. Phosphorus must be applied close to the plant's roots in order for the plant to utilize it. Large applications of phosphorus without adequate levels of zinc can cause a zinc deficiency.

Potassium is necessary for formation of sugars, starches, carbohydrates, protein synthesis and cell division in roots and other parts of the plant. It helps to adjust water balance, improves stem rigidity and cold hardiness, enhances flavor and color on fruit and vegetable crops, increases the oil content of fruits and is important for leafy crops. Deficiencies result in low yields, mottled, spotted or curled leaves, scorched or burned look to leaves..

Sulfur is a structural component of amino acids, proteins, vitamins and enzymes and is essential to produce chlorophyll. It imparts flavor to many vegetables. Deficiencies show as light green leaves. Sulfur is readily lost by leaching from soils and should be applied with a nutrient formula. Some water supplies may contain Sulfur.

Magnesium is a critical structural component of the chlorophyll molecule and is necessary for functioning of plant enzymes to produce carbohydrates, sugars and fats. It is used for fruit and nut formation and essential for germination of seeds. Deficient plants appear chlorotic, show yellowing between veins of older leaves; leaves may droop. Magnesium is leached by watering and must be supplied when feeding. It can be applied as a foliar spray to correct deficiencies.

Calcium activates enzymes, is a structural component of cell walls, influences water movement in cells and is necessary for cell growth and division. Some plants must have calcium to take up nitrogen and other minerals. Calcium is easily leached. Calcium, once deposited in plant tissue, is immobile (non-translocatable) so there must be a constant supply for growth. Deficiency causes stunting of new growth in stems, flowers and roots. Symptoms range from distorted new growth to black spots on leaves and fruit. Yellow leaf margins may also appear.

Micronutrients

Micronutrients are those elements essential for plant growth which are needed in only very small (micro) quantities . These elements are sometimes called minor elements or trace elements, but use of the term micronutrient is encouraged by the American Society of Agronomy and the Soil Science Society of America. The micronutrients are boron (B), copper (Cu), iron (Fe), chloride (Cl), manganese (Mn), molybdenum (Mo) and zinc (Zn). Recycling organic matter such as grass clippings and tree leaves is an excellent way of providing micronutrients (as well as macronutrients) to growing plants.

Iron is necessary for many enzyme functions and as a catalyst for the synthesis of chlorophyll. It is essential for the young growing parts of plants.

Deficiencies are pale leaf color of young leaves followed by yellowing of leaves and large veins. Iron is lost by leaching and is held in the lower portions of the soil structure. Under conditions of high pH (alkaline) iron is rendered unavailable to plants. When soils are alkaline, iron may be abundant but unavailable. Applications of an acid nutrient formula containing iron chelates, held in soluble form, should correct the problem.

Manganese is involved in enzyme activity for photosynthesis, respiration, and nitrogen metabolism. Deficiency in young leaves may show a network of green veins on a light green background similar to an iron deficiency. In the advanced stages the light green parts become white, and leaves are shed. Brownish, black, or grayish spots may appear next to the veins. In neutral or alkaline soils plants often show deficiency symptoms. In highly acid soils, manganese may be available to the extent that it results in toxicity.

Boron is necessary for cell wall formation, membrane integrity, calcium uptake and may aid in the translocation of sugars. Boron affects at least 16 functions in plants. These functions include flowering, pollen germination, fruiting, cell division, water relationships and the movement of hormones. Boron must be available throughout the life of the plant. It is not translocated and is easily leached from soils. Deficiencies kill terminal buds leaving a rosette effect on the plant. Leaves are thick, curled and brittle. Fruits, tubers and roots are discolored, cracked and flecked with brown spots.

Zinc is a component of enzymes or a functional cofactor of a large number of enzymes including auxins (plant growth hormones). It is essential to carbohydrate metabolism, protein synthesis and internodal elongation (stem growth). Deficient plants have mottled leaves with irregular chlorotic areas. Zinc deficiency leads to iron deficiency causing similar symptoms. Deficiency occurs on eroded soils and is least available at a pH range of 5.5 - 7.0. Lowering the pH can render zinc more available to the point of toxicity.

Copper is concentrated in roots of plants and plays a part in nitrogen metabolism. It is a component of several enzymes and may be part of the enzyme systems that use carbohydrates and proteins. Deficiencies cause die back of the shoot tips, and terminal leaves develop brown spots. Copper is bound tightly in organic matter and may be deficient in highly organic soils. It is not readily lost from soil but may often be unavailable. Too much copper can cause toxicity.

Molybdenum is a structural component of the enzyme that reduces nitrates to ammonia. Without it, the synthesis of proteins is blocked and plant growth ceases. Root nodule (nitrogen fixing) bacteria also require it. Seeds may not form completely, and nitrogen deficiency may occur if plants are lacking molybdenum. Deficiency signs are pale green leaves with rolled or cupped margins.

Chlorine is involved in osmosis (movement of water or solutes in cells), the ionic balance necessary for plants to take up mineral elements and in photosynthesis. Deficiency symptoms include wilting, stubby roots, chlorosis (yellowing) and bronzing. Odors in some plants may be decreased. Chloride, the ionic form of chlorine used by plants, is usually found in soluble forms and is lost by leaching. Some plants may show signs of toxicity if levels are too high.

Nickel has just recently won the status as an essential trace element for plants according to the Agricultural Research Service Plant, Soil and Nutrition Laboratory in Ithaca, NY. It is required for the enzyme urease to break down urea to liberate the nitrogen into a usable form for plants. Nickel is required for iron absorption. Seeds need nickel in order to germinate. Plants grown without additional nickel will gradually reach a deficient level at about the time they mature and begin reproductive growth. If nickel is deficient plants may fail to produce viable seeds.

Sodium is involved in osmotic (water movement) and ionic balance in plants.

Cobalt is required for nitrogen fixation in legumes and in root nodules of nonlegumes. The demand for cobalt is much higher for nitrogen fixation than for ammonium nutrition. Deficient levels could result in nitrogen deficiency symptoms.

Silicon is found as a component of cell walls. Plants with supplies of soluble silicon produce stronger, tougher cell walls making them a mechanical barrier to piercing and sucking insects. This significantly enhances plant heat and drought tolerance. Foliar sprays of silicon have also shown benefits reducing populations of aphids on field crops. Tests have also found that silicon can be deposited by the plants at the site of infection by fungus to combat the penetration of the cell walls by the attacking fungus. Improved leaf erectness, stem strength and prevention or depression of iron and manganese toxicity have all been noted as effects from silicon. Silicon has not been determined essential for all plants but may be beneficial for many.

Essential Elements and their Important Functions

Elements	Form in which the element are taken in	Region of the plant that requires the element	Function
Nitrogen, N	NO_2-, NO_3- or NH_4 ions	All tissues, particularly in meristematic tissues	Required for the synthesis of amino acids, protiens nucleic acids, vitamins, hormones, coenzymes, ATP and chorophyII.
Phosphorus, P	H_2PO_4- or HPO_4^{2-}	Young tissues from the older metabolically less active cells	Required for the sythesis of nucleic acids phospholipids, ATP, NAD and NADP Constituent of cell membrane and some proteins.
Potassium, K	K^+	Meristematic tissues buds, leaves and root tips.	Activates enzymes associated with K+ / Na+ pump in active transport, anion-cation balance in the cells. Brings about opening and closing of stomata. Common in cell sap in plant cell vacuole and helps in tugidity of cells.
Calcium Ca	Ca^{2+}	Meristematic and differentiating tissues. Accumulates in older leaves	Present as calcium pectate in the middle lamella of cell walls that joins the adjacent cells together. Activates enzymes needed for the groqth of root and shoot tip. Needed for normal cell wall dvelopment. Required for cell division cell enlargement.
Magnesium Mg	Mg^{2+}	Leaves of the plant	Forms part of the chlorophyII molecule. Activates enzymes of phosphate metabolism. Important for synthesis of DNA and RAN. Essential for binding ofribosome subunits.
Sulphur, S	SO_4^{2-}	Stem and root tips young leaves of the plant	As a constituent of amno acids cysteine and methionine and of some proteins. Present in co-enzyme A, vitamin thiamine, biotin and ferredoxin. Increases root

			development. Increases nodile formation in legumes.
Iron, Fe	Fe^{3+}	Leaves and seeds	Needed for the sythesis ofchlorophyII. As a constituent of ferredoxin and cytochromes.Activates the enzymes catalase.
Manganese Mn	Mn^{2+}	All tissues. Clooects along the leaf veins.	Activates many enzymes of photosynthesis, repiration and N_2 metabolism. Acts as aelectron donor for chloorophyII b. Involved in decarboxylation reactions during respiration.
Molybdenum Mo	MoO_2^{2+}	All tissues particularly in roots.	Requires for nitrogen fixation, Activates the enzyme nitrate reductase
Boron, B	BO_3^{3-} or $B_4O_7^{2-}$	Leaves and seeds	Increases the uptake of water and calcium. Essential for meristem activity and frowth of pollen tube. Involved in translocation of carbohydrates
Copper, Cu	Cu^{2+}	All tissues	Component of oxidase enzymes and plastocyanin. Involved in electron transport in photosynthesis
Zinc, Zn	Zn^{2+}	All tissues	Componene of indoleacetic acid - a plant horm=one. Activates dehydrogenases and carboxylases. Present in enzyme carbonic anhydrase
Chlorine, Cl	Cl^-	All tissues	Essential for oxygen evolution in photosynthesis anion-cation balance in cells.

CHAPTER - 28

Phytohormones

Plant hormones - They are defined as:

1. Small
2. Organic compounds;
3. Synthesized by the plant;
4. Active in low concentration (<10^{-6});
5. Promote or inhibit growth and developmental responses;
6. Often show a separation of the site of production and the site of action (although this isn't as clear a distinction as in animals).

Based on these criteria, would the following substances be considered hormones? Ca^{2+}, sucrose, 2 4 - D, glycine, or K^+?

PLANTS VS. ANIMALS

There are a few significant differences in the nature of hormones found in plants and animals. These are summarized in the following table:

	Plants	Animals
Number of hormones	Fewer	Many
Specificity of action	Non-specific	Specific
Work together	Yes	No
Site action/production separation	No	Yes

Thus, there are comparatively few plant hormones, each elicits a variety of responses and often works together with other hormones. In contrast, animals have numerous different kinds of hormones, each with a specific function, and it works alone to induce a response.

Plant Hormones

There are five major groups, based on chemical structure. With the exception of the latter two, each group represents a family of related compounds. These groups are: (1) auxins; (2) gibberellins; (3) cytokinins; (4) abscisic acid; and (5) ethylene. In addition, there are a variety of other plant "hormones" including the brassinosteroids, oligosaccharides, polyamines, jasmonic acid, salicylic acid, systemin, and putative hormones like those involved in flowering (florigen).

How do we know a substance is a hormone?

In the past, physiologists simply applied the substance to a plant or excised part to see if it caused a response. If there was a response, then we assumed that the substance was a hormone involved in the response. However, responding to an exogenous application of a hormone doesn't necessarily mean that the substance has any endogenous (*in vivo*) action. Thus, we've become a little more picky and only accept that a substance is a hormone if:

1. Exogenous application causes the response;
2. Lowering endogenous levels prevents the response;
3. Lowering endogenous levels followed by exogenous application restores the response;
4. Endogenous levels should be related to the response (i.e., increase *before* the response occurs).

Mechanism of hormone action

Hormones act on target tissues to activate a receptor. The general mechanism is:

hormone α target tissue/cell α receptor α signal amplification à response

Thus, for a response to occur:

1. the hormone must be present in sufficient quantity;
2. the target tissue must be sensitive (sensitized) to the hormone;
3. the target tissue recognizes the hormone (i.e., there must be a receptor to which the hormone can bind);

4. the binding of the hormone/receptor should initiate a change in the receptor (amplification). Calcium is often involved and its interaction is mediated by the protein calmodulin; and
5. the activated receptor initiates a physiological response

Techniques to study hormones

A. Bioassays

A bioassay examines the effect of a test substance on a plant tissue. To perform a hormone bioassay, a test plant is chosen that lacks the hormone for a response. Known amounts of hormone are added to the plant, the response is measured, and a "standard curve" is produced. To determine if a sample contains the hormone, the test plant is treated in a similar fashion. If present the hormone can be quantified by comparing its response to the samples of known concentration. For example, a variety of rice (Tauginbozu) lacks GA and is often used in bioassay studies. After treatment with various concentrations of GA, leaf length vs. [GA] is plotted.

Table: Comparison of the advantages and disadvantages of bioassays

Advantages	Disadvantages
Simple & easy inexpensive	Sensitive to impurities false positive tests can result lower sensitivity than other methods

B. Immunological studies

Antibodies are made against the plant hormones and then used as specific probes to localize and quantify. The antibodies are usually coupled to radioisotopes or fluorescent dyes to make it easier to trace. This technique is very sensitivity and specific.

C. Instrumental Methods: GC-MS; HPLC; high specificity and sensitivity

Methods for regulating endogenous levels

The internal levels of plant hormones must be tightly controlled so that the response occurs only at the appropriate time. Regulation of hormonal action is achieved by: (1) controlling the rate of hormone synthesis; (2) forming conjugates, which are inactive storage forms where the hormone is covalently bonded to a sugar, amino acid or other molecule; (3) enzyme degradation; (4) transporting the hormone away/toward the site; and (5) compartmentalizing the substance in an organelle such as the chloroplast.

Numerous synthetically produced growth regulators display hormone-like effects. They have a decisive economic importance as herbicides or growth stimulators in modern agriculture and horticulture, and - due to their dangerousness and the toxicity of their by-products (dioxin!) - an explosive political potential. Some important phytohormones covered here are auxins, cytokinins, gibberellins, abscisic acid and ethylene.

1. AUXINS

Auxin is a general name for a group of hormones that are involved with growth responses (i.e., elongate cells, stimulate cell division in callus). Not surprisingly, the term "auxin" is derived from the Greek word "to increase or grow". This was the first group of plant hormones discovered.

The existence of auxins was realized as early as in 1887, when Charles Darwin and his son Francis Darwin demonstrated that canary grass coleoptiles did not show phototropic response when the tip was either removed or covered with a foil tip. Now we know that auxins are present in the tissues of all higher plants, although in small quantities. They are of fundamental importance in the physiology of growth and differentiation.

Chemical structure

The first crystalline auxin was obtained from human urine by Kogl and others in 1934. Chemically it was shown to be indole 3-acetic acid(IAA) Since than it has been found that IAA occurs as a principal auxin in most of plant species.. The formula shows that it is a tryptophane derivative.

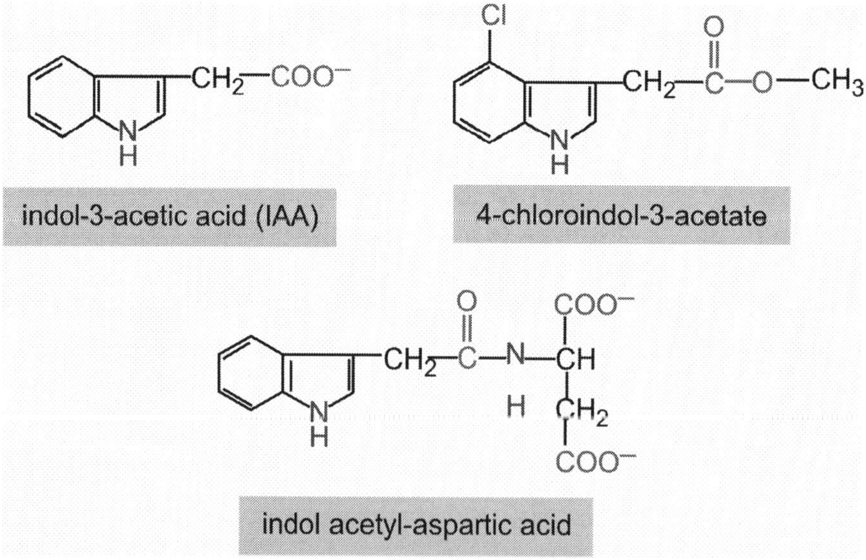

indol-3-acetic acid (IAA)

4-chloroindol-3-acetate

indol acetyl-aspartic acid

It turned out that auxin is a collective name for several similar compounds. Auxin occurs in cells in concentrations of 10-8 – 10-6 Mol/l. As we know today are IES and its similar compounds very common in green plants and fungi. Methyl-4-chlorindole-3-acetic acid and indole aspartate, for example, were found in unripe pea seeds (and unripe seeds of other plants, too). Auxins are often glycosylated or bound to proteins.

Biosynthesis

A. Site

Auxin is made in actively growing tissue which includes young leaves, fruits, and especially the shoot apex. Made in cytosol of cells

B. Routes

There are two major routes to the production of IAA.

1. Tryptophan-dependent Pathways. The similarity of chemical structure of IAA and tryptophan suggested a connection between these. Considerable research has shown that tryptophan, one of the protein amino acids, is a precursor of auxin biosynthesis. Overall, the conversion of tryptophan to IAA can occur by: (1) a transamination followed by a decarboxylation; (2) a decarboxylation followed by a transamination; or (3) formation of IAA via an oxime (C=NOH) and nitrile (CN).
2. Tryptophan-independent Pathway - this route doesn't involve tryptophan directly as an intermediate to the formation of auxin.

Transport

A. Basipetal (or Polar) transport

Auxin is transported in a basipetal (towards the base, base-seeking) direction. In other words, auxin moves from the shoot tip towards the roots and from the root tip towards the shoot.

Evidence - (1) Seedling vs. [auxin]; and (2) ^{14}C labeled IAA applied to the top of a stem section is recovered only from the bottom of the stem section. When auxin is applied to the end of stem segments, it is only transported from the "top" of the section to the "bottom" as demonstrated in these data:

	Upright	Upside down
^{14}C-IAA in donor block (dpm)C-IAA in donor block (dpm)	10,000	10,000
^{14}C-IAA in receiver block (dpm)C-IAA in receiver block (dpm)	7,500	300

B. Mechanism of polar transport

1. Transport Rate – IAA is not transported through the transpiration stream or phloem because the rate of movement is too slow. The rate of transport is consistent with diffusion.
2. Tissue of transport - appears to occur in parenchyma cells associated with the vascular tissue.
3. Model for polar transport: (a) Protons are moved out of the cell by a proton pump that requires ATP; (b) IAA is protonated at low pH; (c) IAA-H passes through the lipid membrane. It can enter or leave anywhere; (d) once inside the cell, the IAA-H ionizes in response to the higher pH; (e) IAA^- requires a permease to pass through the membrane; (f) histochemical studies have shown that a permease is only located at the bottom of the cell - resulting in a net movement of auxin out of the bottom of the cells.
4. Evidence for polar transport model: (a) transport is blocked by respiratory poisons (i.e., demonstrates the need for ATP and a proton pump); (b) unlabeled IAA competes with C^{14} labeled IAA for uptake in to the cells - this suggests that a carrier of some sort is required ; (c) fluorescein-labeled antibodies show that the permease is localized at bottom of cells; and (d) the transport of auxin is blocked by NPA (napthylthalamic acid), TIBA (tri-iodobenzoic acid) and flavanoids. These inhibitors appear to block the permease.
5. Proton-Auxin CoTransport Mechanism - In addition to the passive pH-dependent mechanism described, there a membrane transport protein seems involved that cotransports protons and IAA into the cell. This tranport protein is localized in the upper side of the cells.

Bioassay

Detection of a plant growth substance using a biological test material is termed as bioassay. The most widely used bioassay of auxins is the Avena coleoptile curvature test, which has been used since early works on auxins. Unilateral application of solution suspected to contain auxin causes curvature in the coleoptile.

Physiological effects

1. **Stem elongation** : Application of exogenous auxins causes stimulation of coleoptile and stem growth. At high concentrations (say above 10-4M), auxin inhibit the growth. This inhibition is believed to be associated with ethylene accumulation at higher auxin concentration.

2. **Coleoptile curvature effect :** Bending occurs because of the polar transport of auxin. The presence of auxins on only one side of coleoptiles increases cell division on that side and therefore the coleoptiles bends on other side.
3. **Apical dominance :** The inhibition of lateral bud formation by the presence of apex is called apical dominance. This inhibition is attributed to the presence of auxins in the apex.
4. **Rooting :** The auxin IAA has been identified as a rooting hormone. Exogenous application of IAA induces rooting in several stem cuttings. Elongation of roots also increases with the application of exogenous auxins although in most cases it inhibits root growth in intact plant.
5. **Cambial activity** : Stimulation of cambial activity in trees by exogenous application of auxins was first reported in 1936 by Soding.
6. **Flowering :** The exogenous application of auxin increases formation of female flowers and ovary wall growth leading to parthenocarpy in cucurbits, grapes etc.
7. **Abscission :** C.D. Laure(1936) demonstrated for the first that many synthetic auxins inhibited the abscission of coleus leaves. Since then control of abscission by IAA has been demonstration in various other plants as well. In some cases IAA induces abscission, especially in the aged tissues.
8. **Plant growth movements :** Because of their polar transport and effect on cellular growth, auxins are known to be involved in photropic and geotropic lant movements.

Mechanism of action

Auxins affect a variety of physiological processes in plants. Some of these effects are very fast, while others have a lag of few minutes to few hours. It appears however, that at molecular level and perhaps other plant hormones as well, affect nucleic acid directed protein synthesis, enzyme activity and membrane permeability, either directly or through some other metabolites. All physiological effects are manifestations of one or more of these effects.

2. GIBBERELLINS

- Associated with the fungus, *Gibberella fujikuroi*.
- liquid from the culture medium applied to plants caused symptoms (Kurosawa, 1926).

- three gibberellins were isolated and identified from the medium & fungus
- independent confirmation in labs in Europe and US.
- gibberellins known from angiosperms, gymnosperms, ferns, mosses, algae, fungi and even a few bacteria.
- Today are more than 110 different gibberellins known (GA1, GA2,....GA3, GA4.....GA110) that differ only little chemically but very much in their biological activities.
- Roughly 30 percent of all known gibberellins are biologically active.
- All higher plants contain presumably at least one, but usually several active and inactive gibberellins that exist in different concentrations depending on the respective tissue.

Chemical structure

Gibberellins are diterpenoids derived from four isoprenoid units forming a system of four rings. It is distinguished between gibberellins of 19 and such of 20 C-atoms. The twentieth C-atom is not part of the four rings but belongs to a side chain (CH3 in GA12, CH2OH in GA15, CHO in GA19 or COOH in GA28). Both enumeration and the illustration below show, how the single structures differ.

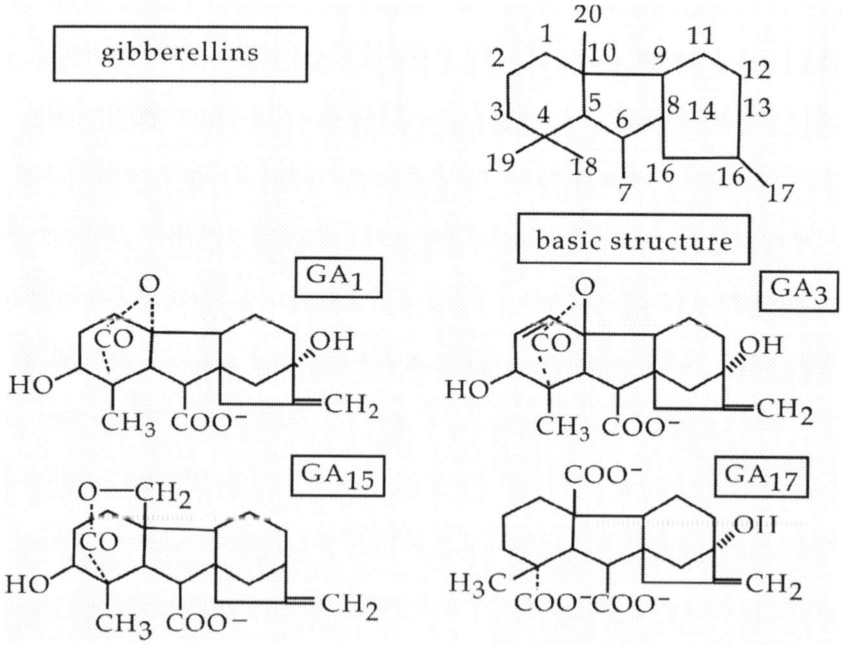

Chemistry

- diterpenes (C20, though some have lost a carbon and are C19) – based on isoprene skeleton
- characterized by having a complex system of 4-5 rings (ent-kaurene ring system) and a carboxyl (acidic) side chain
- more than 110 different gibberellins are known - abbreviated $GA_1...GA_n$.
- No more than 12 or so different gibberellins occur in any one species
- only a few (about 15) of the GA's have biological activity; the rest are likely breakdown products of, or precursors to, the active ones
- GA_1 is the most active GA
- GA_3 was the first one discovered and is readily available commercially (produced by *G. fujikuroi* cultures).
- C19 GA's are more active than C20
- GA's can occur in a free or conjugated form (i.e., glucosides)

Biosynthesis

A. Site - young leaves, roots, and developing seeds (developing endosperm) and fruits.

B. Pathway

- terpene pathway
- basic terpene building block is isoprene (isopentenylpryophosphate, IPP), a five carbon unit
- five major steps: (1) synthesis of IPP; (2) condensation of 4 isoprene units to form geranylgeranylpyrophosphate (GGPP); (3) GGPP cyclizes to form the kaurene ring system; (4) a methyl group of kaurene is oxidized to a carboxyl group to form GA_{12}-aldehyde; (5) GA_{12}-aldehyde is the precursor to the other GA's
- two routes to the production of IPP (isoprene): (1) Mevalonate-dependent pathway - occurs in cytosol; mevalonate is especially important for sterol biosynthesis; mevalonate is derived from acetyl CoA; and (2) Mevalonate independent pathway – especially in plastids, important for carotenoid biosynthesis

C. GA synthesis inhibitors

- Kaurene synthesis inhibitors - Phosphon D, CCC (cycocel), and Amo1618.
- Kaurene oxidation inhibitors - Ancymidol (A-rest) and paclobutural
- Another inhibitor is B-Nine (alar).

Transport

- made in the tissue in which it is used
- transport occurs through xylem, phloem, or cell-to-cell.
- phloem seems to be most important transport route
- transport is not polar, as it is for auxin.
- not purely diffusion.

Disposal/Regulation of endogenous levels

- Formation of conjugates
- Slow hydrolysis

Bioassays/Analysis

A. Three common bioassays used for gibberellin are:

- Lettuce hypocotyl elongation
- Dwarf rice (var. Tanginbozu rice) leaf sheath elongation
- alpha-amylase production in barley

B. Instrumental methods – now the method of choice, especially GC-MS

Physiological effects

1. **Promotion of plant growth :** Gibberellins enhance elongation of intact stems much more than that of excised stem segments. Because of this property gibberellins are able to overcome genetic dwarfness in some species like maize, bean, pea etc.
2. **Promotion of seed germination and bud growth :** Gibberellins promote seed germination in several species, which otherwise fail to germinate unless subjected to low temperature, long days or red light. They enhance cell elongation so the radicle can push through

endosperm, seed coat or fruit that restricts its growth. They also overcome dormancy in buds of evergreen and deciduous trees and shrubs.

3. **Induction of flowering :** Flowering in many herbaceous plants is correlated with rise in endogenous GA level. Exogenous application of gibberellins induces flowering in some photoperiod sensitive and cold requiring species.
4. **Prevention of senescence :** The exogenous application of GA can prevent senescence of leaves. In leaves of *Taraxanum*, GA acts powerfully on senescence.
5. **Mobilization of food and minerals in seeds :** Gibberellins are known to stimulate the hydrolysis of stored macromolecules and their transport to the embryonic axis.

Mechanism of action

Gibberellis seem to affect plant growth and development in a variety of ways. In many cases, it has been found that the GA stimulation of plant growth in dwarf seedlings is associated with increase in auxin content. Further, GA suppresses the activity of IAA oxidase, the enzyme responsible for IAA degradation. It appears that most of the physiological effects of GA are mediated through increased enzyme synthesis and membrane permeability. These aspects are (a) Enzyme induction, (b) Membrane permeability, (c) Gibberellin interaction with DNA and (d) Analogy with animal hormones.

3. CYTOKININS

- Called "cytokinins" because they stimulate cell division (*i.e.*, cytokinesis)
- Haberlandt (1913) noted that non-dividing potato parenchyma cells would revert to actively dividing ones in the presence of phloem sap. This observation suggested a soluble material was responsible for cell division.
- Folke Skoog (1940's) and colleagues at Univ. of Wisconsin found that cultured tobacco pith tissue explants would proliferate only if they were supplemented with various substances such as autoclaved herring sperm or coconut milk.
- Miller (1956) identified the first cytokinin, called kinetin, in the herring sperm.

Chemical structure

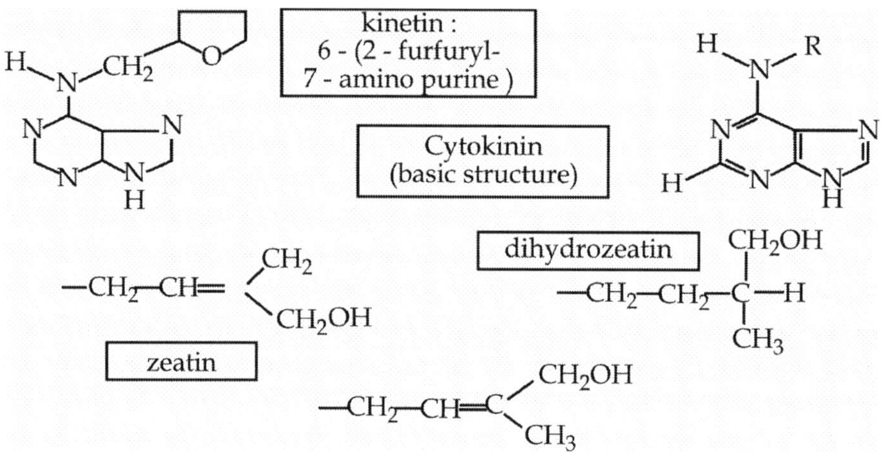

kinetin :
6 - (2 - furfuryl-
7 - amino purine)

Cytokinin
(basic structure)

dihydrozeatin

zeatin

Most of the endogenous cytokinins are purine (adenine) or purine derivatives. Zeatine, the most active of naturally occurring cytokinins is 6- amino purine(2-furfurul-7 amino purine). Zeatine exists normally as ribotide(base + ribose + phosphate) derivatives and as such it is present in RNA specially in the t- RNA of many higher plants and bacteria

Chemistry

A. General

- Adenine derivatives (amino purines)
- Occur as: (a) the free nitrogenous base; (b) a nucleoside (base + ribose); (c) a nucleotide (base + ribose + phosphate); or (d) glycosides
- The free base is the active form.
- Approximately 40 different structures known.
- Zeatin (Z), which was first isolated from maize (*Zea mays*) is the most common cytokinin.
- Other naturally occurring cytokinins include, dihydrozeatin (DHZ) and isopentenyladenosine (IPA).

B. Synthetic cytokinins

- Kinetin (*as above*) – probably byproduct of zeatin degradation
- There are several other substances with cytokinin activity such as benzyl adenine (benzylaminopurine; BA).

C. Cytokinins and nucleic acids

- Can occur as a modified base in tRNA, but the bases exist in the *cis* form, rather than the typical *trans* form. These modified bases that are found in all organisms from bacteria to plants to humans.
- The function of the tRNA cytokinins is not clear, but after hydrolysis of the tRNA the products can act as a cytokinin. The importance of the tRNA derived cytokinins in overall growth and development is not clear, either.
- Interestingly plants have different sets of tRNA's with different cytokinins that participate in protein synthesis in the cytoplasm and the plastids.

Synthesis

- Site: synthesized primarily in the meristematic region of the roots. This is known in part because roots can be cultured (grown in artificial medium in a flask) without added cytokinin, but stem cells cannot.
- Cytokinins are also produced in developing embryos and crown gall tissues
- First major precursor to the cytokinin is AMP (adenosine monophosphate)
- Side chains of the cytokinins are made by the terpene pathway (IPP - isoprene) and added to the AMP by cytokinin synthase
- There is some speculation that surface and endophytic bacteria (*i.e., Methylobacterium*) may be the actual source of plant cytokinins. Rationale: (1) haven't isolated some of the putative genes for cytokinin synthesis; (2) remove bacteria show impaired cytokinin production.
- tRNA nucleotides are modified to cytokinins after the tRNA is transcribed (post-transcriptional processing)

Transport

- Via xylem (transpiration stream)
- In peas, a signal from the leaves may signal/regulate transport of cytokinins from the roots
- Zeatin ribosides are the main transport form; converted to the free base or glucosides in the leaves
- Some cytokinin also moves in the phloem.

Bioassays/Analysis

A. Bioassay
   1. Callus culture cell proliferation - not used too much because it takes too long
   2. Expansion of radish or cocklebur cotyledons
   3. Inhibition of chlorophyll loss by detached oat leaves during senescence

B. Methods of Analysis – liquid chromatography, mass spectroscopy; radioimmunoassays

Disposal

- Forms conjugates with glucosides. major storage form of cytokinin, inactive
- Cytokinin oxidase may be an important route of disposal, too.

Bioassay

The most sensitive and specific bioaasay is plant callus culture. Another bioassay of cytokinins is the lettuce seed germination test. However this bioassay is not very specific, as some gibberellins are also known to promote lettuce seed germination in dark. Prevention of xanthium and barley leaf senescence has also been used as bioassay of cytokinins.

Physiological effects

1. **Promotion of cell division and organ formation :** A major function of naturally occurring cytokinins is to promote cell division. In the presence of auxins, nearly all cytokinins stimulate cell division and subsequent callus growth in the parenchymatous cells from several plants. By manipulating the ratio of cytokinin to auxin, plantlets can be developed from the tissue in the culture.
2. **Promotion of seed germination:** Application of cytokinin can promote germination and even break the dormancy in some seeds. Lettuce seeds are responsive to cytokinins. If the seeds are soaked in a solution of a cytokinin, their germination is greatly enhanced in the dark.
3. **Expansion of cotyledons and leaves :** Cytokinins cause cotyledon expansion by increasing cell size. About 10^{-6} M benzyl adenine can increase size of *xanthium* cotyledons by almost five times. Cytokinins also stimulate the growth of leaf and also cause the expansion of leaf

disc in some dicots. 50μM kinetin causes 2.5 times increase in the diameter of bean leaf discs during 48h treatment in dark.

4. **Delaying of senescence :** Cytokinin mediated delaying of senescence is linked with its effects on translocation of metabolites. Cytokinin treated leaves withdraw nutrients from the adjoining leaves. Thus, while they are able to maintain their nutrition and vigour, the adjoining leaves begin to senesce first.
5. **Promotion of chloroplast development :** Exogenously applied cytokinins promote chloroplast developments in callus tissues and excised cotyledons. Application of cytokinin in dark promotes lamellar development and if light is also applied simultaneously grana and chlorophyll also appear.
6. **Anthocyanin synthesis :** Cytokinin treatment increases anthocyanin content in many cultured cells and tissues and in parts of intact plants. For example in Arabidopsis seedlings, about 50 fold increase in anthocyanin content has been observed at 0.5μM benzyladenine.

Mechanism of action

The most important aspect of cytokinin action is increased mitosis which involves increased synthesis of nucleic acids and proteins. However, some other physiological effects are not easily explained through such a mechanism. Some possible mechanisms of cytokinin action are (a) Increased nucleic acid and protein synthesis (b) Increased enzyme activity and (c) Involvement of a receptor protein and a second messenger.

4. ABSCISIC ACID (ABA)

- Bennet-Clark and Kefford (1953) described the presence of an inhibitor of coleoptile elongation in oats
- About 10 years later, Addicott *et al.* found a substance that stimulated abscission of fruits in cotton and they named it abscisin II
- About the same time, Wareing found a substance in sycamore leaves that promoted dormancy in buds and called it dormin
- It was soon clear that these were the same substance
- A conference in 1967 straightened out the name and it was decided to call the hormone abscisic acid (ABA).

Chemical structure

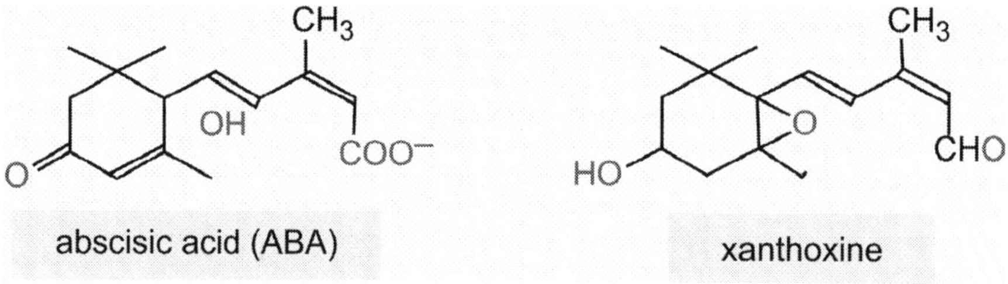

abscisic acid (ABA)

xanthoxine

In some plant tissues (especially in young shoots) occurs a related compound called xanthoxine. Whether xanthoxine is an intermediate of the ABA-biosynthesis or whether it is an independent product remains unknown. The structure indicates that both ABA and xanthoxine are terpene derivatives. This was proven when it could be shown that radioactively labelled mevalonic acid is integrated into ABA though it does not elucidate which intermediates are produced. Two alternative biosyntheses have been discussed:

ABA is a degradation product of xanthophyll (especially of violaxanthin).

ABA is produced from a C15 precursor using a separate pathway and is thus independent from the carotenoid/xanthophyll metabolism.

The first idea seemed initially more plausible since the structures of xanthophylls and ABA correspond to a large degree.

Chemistry

- A single structure, not a family of related structures like the gibberellins
- Sesquiterpene (*i.e.*, terpenoid) - C15 - made from 3 isoprene units
- Occurs as *cis* form; S-enantiomer is natural and active form
- Found in all green plants, also in some mosses, algae, and fungi
- Related to lunularic acid which is found in liverworts.

Biosynthesis

- Plastids
- Most tissues, especially leaves and seeds
- Terpene pathway
- IPP is the first intermediate; not derived from mevalonate (mevalonate-independent); comes from intermediates of glycolysis

- ABA derived from the breakdown of carotenoids. Evidence: maize mutants blocked in carotene synthesis ❼ low levels of ABA ❼ vivipary (see below)
- Pathway: IPP ❼ farnesyl pyrophosphate ❼ carotenoids (C40; violaxanthin) ❼ xanthoxin ❼ ABA

Bioassays/Analysis

A. Bioassays – there are several including:
   - inhibition of seed germination
   - inhibition of GA induced alpha-amylase production

B. Analysis – Gas chromatography, HPLC, and immunoassay

Disposal/Regulation

- Rapid changes in endogenous levels (up to 100x within a few days)
- ABA levels can be regulated by: (a) degradation; (b) compartmentalization; (c) transport; (d) conjugation to a sugar or other molecule; and (e) conversion (oxidation) into phaseic acid and dihydrophaseic acid.

Transport

Xylem and phloem (greater amounts)

Bioassay

Many bioassays have been developed for detecting ABA. Inhibition of seed germination is often used to detect the presence of ABA in plant tissues. Another bioassay is the test on elongation growth of seedlings. In some cases about 1mM ABA can cause as high as 50% inhibition of the elongation growth in 2 hours. Other bioassays of ABA include inhibition of α-amylase synthesis, stomatal closing in epidermal peels and acceleration of abscission in excised tissues.

Physiological effects

1. **Inhibition of seed germination:** Exogenous supply of ABA inhibits germination of most non-dormant seeds. For causing an inhibition, ABA must be continuously present. Endogenous ABA also inhibits germination.

2. **Inhibition of seedling growth:** In *Glycine max for example,* 1Mm ABA inhibits seedling growth by about 50% in 48 hours. In some cases, similar growth inhibition is observed in almost 2 hours. ABA inhibits seedling growth by decreasing the water potential.
3. **Stomatal closing:** The most significant and best known effect of ABA is its control of stomatal closing. Its exogenous application to epidermal strips causes stomatal closure. The response of ABA on stomatal closure is very fast and occurs within a few minutes of its application.
4. **Geotropism:** The accumulation of ABA in the tip appears to require light and gravity. ABA produced in the cap seems to be translocated basipetally and to stimulate a positive geotropic response. Further, exogenously applied ABA induces positive geotropism although it inhibits root growth.
5. **Senescence and abscission:** ABA is an endogenous factor involved in senescence and abscission of leaves and other plant organs. Exogenous application of ABA induces primary yellowing in lesf tissues in a variety of species ranging from deciduous to herbaceous plants.

Mechanism of action

Mechanisms of action of ABA include (a) Membrane permeability, (b) nucleic acid directed protein synthesis and (c) Involvement of a secondary messenger.

5. ETHYLENE

- The Chinese may have been the first to observe the effects of ethylene when they noted that burning incense increased fruit ripening
- In 1864 leaks in gas lights in street lamps were reported to stunt plant growth and defoliate trees
- In 1901, D. Neljubow realized that his dark-grown pea seedlings were short, fat and negatively gravitropic (the triple response) because of a component in "laboratory air" which he subsequently identified as ethylene
- Cousins (1910) first reported that ethylene occurred in plants.

Chemical structure

Ethylene is a gaseous effector with a very simple structure. Nonetheless, ethylene has features that identify it as a hormone such as the fact that it is effective at nanomolar concentrations.

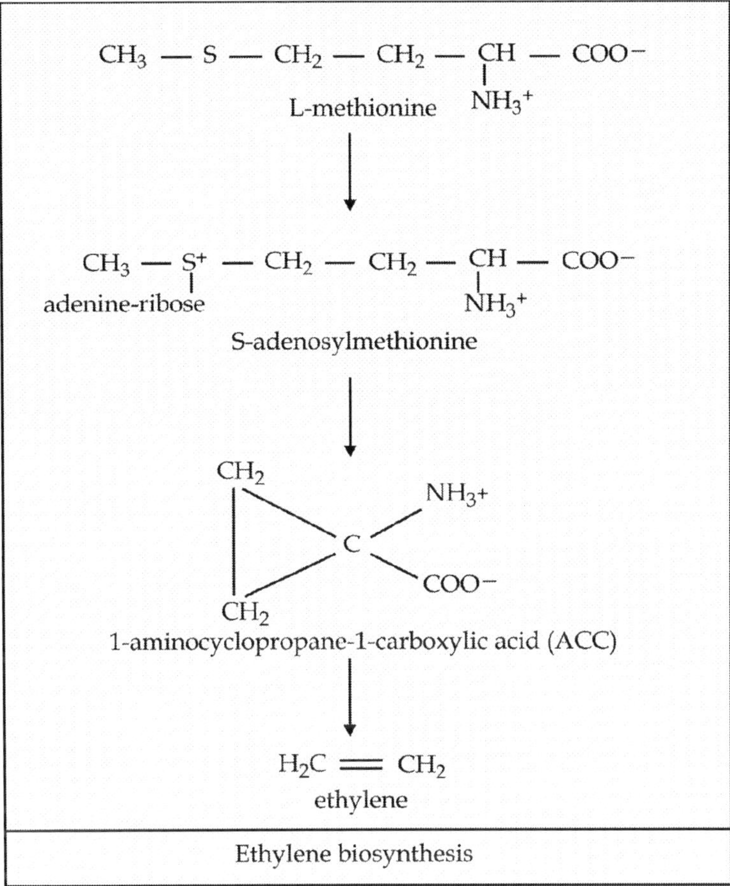

Ethylene biosynthesis

Chemistry

- Single compound (like ABA) and is not a family of related ones (*i.e.*, gibberellins)
- CH_2=CH_2
- Ethylene (MW 28) is similar in size/shape as water
- A gaseous plant hormone

Biosynthesis

A. General:

- Made by most plants including angiosperms, gymnosperms, ferns, mosses, liverworts
- Also synthesized by fungi and bacteria

- Made by all parts of the plant
- Meristematic regions (shoot apex) and senescing tissues are rich sources
- Nodes make more ethylene than internodes
- Ethylene production is stimulated by physiological stresses including wounding, anaerobic conditions, flooding, chilling, disease and drought.
- During the climacteric – which is the sudden surge of respiratory activity that occurs at the peak of ripening in many fruits - lots of ethylene is made

Pathway of synthesis

- The first precursor is methionine (one of the protein amino acids)
- Methionine + ATP $\rightarrow$ S-adenosyl-methionine (SAM) $\rightarrow$ ACC synthase $\rightarrow$ amino cyclopropane carboxylic acid (ACC) $\rightarrow$ ACC oxidase $\rightarrow$ $CH_2=CH_2$ + HCN (hydrogen cyanide) + CO_2. To summarize the highlights:

  1. ATP reacts with methionine to form SAM
  2. SAM is essentially a carrier form of methionine (it is involved in other reactions in the cell)
  3. ACC synthase is the most crucial enzyme in the pathway. It is the rate limiting step and induced by: (a) fruit ripening; (b) flower senescence; (c) in response to IAA; (d) under wounding; (e) chilling injury; (f) drought; (g) flooding; (h) in response to ethylene ("one bad apple spoils the whole bunch"). ACC synthase is cytosolic and coded by a multi-gene family.
  4. methionine is re-claimed via the Yang cycle.
  5. ACC oxidase was formerly called "ethylene forming enzyme, abbreviated EFE." It requires Fe^{2+} and ascorbate for activity – which explains why it took awhile to characterize this enzyme. It is also the product of a multigene family.
  6. One of the products of ACC oxidase activity is HCN – which explains why most plants have enzyme systems for detoxifying/ metabolizing cyanide. ACC oxidase is inhibited by anaerobic conditions and cobalt ions but stimulated by ripening.

Inhibitors

- Silver ions (Ag^+), CO_2 and $KMnO_4$ inhibit ethylene actions. These bind to ethylene receptors or otherwise interfere with the mechanism of ethylene action.
- Aminovinylglycine (AVG) and aminooxyacetic acid (AOA) block the action of ACC synthase. Since these compounds are knows to block pyridoxal enzymes, it suggested that they participate in the process.

Disposal

- Probably not a concern since it is a gas
- In addition, malonate will bind to ACC forming N-malonyl ACC. Unlike most conjugates that can be hydrolyzed to produce the original compound, this one is NOT converted back to ethylene.
- A conjugate with glutamic acid (GACC) may be an important regulator of ethylene levels

Bioassay

Because of its gaseous nature, ethylene is generally detected by gas chromatography. It can also be measured by manometric method. Detection of ethylene using plant materials as tests, is also used in some investigations.

Physiological effects

1. **Ripening of fruits:** Acceleration of fruit ripening was the first discovered effect of ethylene. The hormone is now known to accelerate ripening of mature fruits in most cases including banana, apple, tomato, avocado etc.
2. **Acceleration of senescence and abscission:** Ethylene is an important hormone governing the senescence and abscission of plant parts both natural and induced. Abscission is the most widely demonstrated response of ethylene.
3. **Seed germination:** Ethylene is known to break dormancy and induce germination of lettuce, groundnut, wheat, clover and cocklebur seeds. It also causes the increased extension growth of the seedling in cocklebur. The extension pf growth is maximum at 0.3ppm ethylene.
4. **Growth inhibition and morphogenetic effects:** In most of cases exogenous application inhibits plant growth. In most dicots the elongation growth of stem, roots and leaves is inhibited. Some growth

promoting effects of ethylene are also known. The hormone appears to be involved in the increased coleoptile and shoot growth of submerged plants such as rice.

5. **Flowering inhibition and sex expression:** Ethylene inhibits flowering in most plants, although it is known to promote flowering in mango and pine.Ethylene changes sex expression also in unisexual plants. It increases female flowers in several members of cucurbitaceae. It also induces male sterility in cucurbits and wheat.

Mechanism of action

Mechanisms of action of ethylene include (a) involvement as a receptor, (b) increased membrane permeability,(c) activation of gene expression and (d) involvement of auxins.

Commercial applications

Ethrel (Ethephon) liquid sprayed onto plants. It contains a dilute solution of 2-chloroethylphosphonic acid that breaks down to give off ethylene. Among others things, it is used to synchronize flowering and fruit set in pineapples. Commercial fruits are usually stored in low O_2 to inhibit ethylene biosynthesis or under high CO_2 to prevent ethylene action as a ripening promoter.

Zeitfracht Medien GmbH
Ferdinand-Jühlke-Straße 7
99095 Erfurt, Deutschland
produktsicherheit@kolibri360.de